SpringerBriefs in Public Health

Miguel Kottow

From Justice to Protection

A Proposal for Public Health Bioethics

 Springer

Miguel Kottow
Public Health School
Faculty of Medicine
Universidad de Chile
Universidad Diego Portales
Santiago, Chile
mkottow@gmail.com

ISSN 2192-3698   e-ISSN 2192-3701
ISBN 978-1-4614-2025-5   e-ISBN 978-1-4614-2026-2
DOI 10.1007/978-1-4614-2026-2
Springer New York Dordrecht Heidelberg London

Library of Congress Control Number: 2011943293

Printed on acid-free paper

Springer is part of Springer Science+Business Media (www.springer.com)

Contents

Chapter 1
On Justice and Health Inequities

Abstract Globalized integration in trade and finances has required political adjustments that drive traditional public services to become commodities subject to market conditions. Most nations have reduced their commitments to provide publicly funded health, education, and social securities, all of which are being progressively privatized. This trend has deepened the gap between rich and poor, leading to increasing social and health inequalities that are impervious to traditional public health policies. Summit meetings and international declarations have acknowledged the need to fund external aid to underdeveloped populations, academia lavishly proposing global justice and health equity based on vaguely conceived pleas for solidarity and fraternity. The appealing ideas of global ethics have clashed with economic realities—*Realpolitik*—as international aid dwindles beyond expectations and promises, while theoretical schemes are relegated to the realm of utopia.

Health and health care gradients and disparities are on the rise, fueled by an epidemiological turn from infectious to chronic degenerative diseases that are associated with socioeconomic determinants and aggravated by man-made risks, such as accidents, ecological deterioration, violence, health disrupting life styles, and unwholesome individual behavior. The reassessment of cultural and personal values indicates the need for applied ethical perspectives to become incorporated into the deliberations of public policies, thus bringing to the forefront the development of bioethics in public health.

Keywords Bioethics • Epidemiological transition • Health inequities • Justice • Public health • Socioeconomic determinants

M. Kottow, *From Justice to Protection: A Proposal for Public Health Bioethics*,
SpringerBriefs in Public Health 1, DOI 10.1007/978-1-4614-2026-2_1,
© Miguel Kottow 2012

Social Justice

Reminded of its brotherhood with politics in forming what used to be called practical philosophy, ethics had been severely neglected in academic scholarship, until it was called upon to try and make sense of modern complexity, its meanings and the values it harbors or has asphyxiated. After the war-torn bloody first half of the twentieth century, a pragmatic turn was required in quest of orientation and guide-lines for action in the maze of multicultural modern societies and the emergence of global scenarios in economy, politics, ecology, science, biomedical disciplines, and social services. A striking, though belated, realization has been that the glow of modernity casts selective shadows on humanity: while some societies bask in postindustrial well-being, many others are exiled into endless misery and hopeless deprivation.

There is ample evidence that inequities between rich nations and poor populations are increasing, the Gini index as a marker of income differences within a society getting worse in countries where disparities have been endemic. The income gradient between nations is also on the rise. How effervescent economic growth affects income disparities is hard to understand, especially when learning that "literally hundreds of scholarly papers on convergence or divergence of countries' income have been published in the last decade based on what we know now were faulty numbers." When corrected, statistics show a gloomy picture as far as worldwide inequality is concerned, with conditions continually worsening (Milanovic 2008). Global inequality rose from an estimate of 65 Gini points to a corrected level of 70 Gini points. In 1992, The United Nations Development Program reported that the richest quintil of the world population received 82.7% of the world GDP, whereas the poorest quintil barely got 0.2%. Remedial policies have been timid, lacking in commitment and ignoring international agreements to provide external aid and encourage development of poor populations.

Though insufficiently heeded by justice theorists, first warnings have been issued that justice is a context-bound social practice, not a universal, self-contained Platonic idea. The preeminence of individual autonomy has arguably been the most severe challenge to the ideal of justice, forcing it to fully enter into the arena of practical reasoning where prudence requires deliberation and specification of maxims that cannot claim absolute value. Inflexible justice would constrain autonomy, much as unfettered autonomy tends to ignore justice. Should justice continue to inspire scholarly attention, it must resist being evicted into the realm of utopia, and face the real world by honoring its current denomination as "social justice." Social justice tends to attach "primary significance to the final allocation of resources among individuals, and only secondary significance to the mechanisms used to arrive at that allocation," giving precedence to substantive justice of procedure (Miller 1999). Social processes stand in the way of pure moral principles and submit them to a *prima facie*—valid unless trumped by an equally valid moral claim—or a *pro tanto*—valid in a given context—consideration. "You cannot just jump out of the human condition," just as you cannot proclaim an absolute justice that abstracts from social practices, human inclinations, and the realities of ordinary life (Taylor 1994). Elaborate discourse too often forgets that "the cart of procedural justice ought not to be put before the horse of substantive justice" (Elster 1989 quoted in Miller 1999).

Globalization

International organizations and philosophers voice the need for a new approach to the issue of justice. Even though interest in justice as an ideal principle has been refocused as social justice, there remains a strong tendency to pose justice in more universal terms in consonance with the globalizing trend of modernity. The universalization discourse developed in the initial phases of globalization was pervaded by the optimistic view that the world will change for the better, originating new opportunities and wealth for all and leading, eventually, to global equality. The tenacity with which institutions and scholars cling to the unrealized idea of global ethics clashes with the realities of globalization These matters need to be addressed here in view of their negative influence on public health, considering that throughout history social justice has never been realized in any society, much less at a global level.

Globalization is a development of late modernity with so wide ranging and diverse effects, that a comprehensive overview is as good as impossible. Respecting the Weberian categories of teleological rationality—*Zweckrationalität*—and a rationality based on beliefs and values—*Wertrationalität*—Habermas distinguishes between cognitive-instrumental rationality that deals with science, technology and humankind's adaptation, and moral-political communicative rationality aimed at man's emancipation. Globalization is the dutiful daughter of an instrumental rationality oblivious to nonpragmatic views.

Essential to the well-oiled functioning of globalization is that all participating States—and hardly any State is audacious enough to resist integration—abandon national policies in economic matters, adopt international tariff agreements, and untie the knots of excessive labor force protection in order to secure cheap and competitive production. Globalization removes "barriers between national borders in order to facilitate the flow of goods, capital, service and labour," as defined by the United Nations Economic and Social Commission for Western Asia.

The idea of a globalized form of democracy loses much of its appeal when it stresses increased representation instead of favoring a reliable structure that will respond to citizens' interests and needs, and neglects to realize how many national politics have failed to actually provide democratic power relations (Weinstock 2006). As States are weakened and become increasingly unable to establish social security policies for their societies, power and wealth disparities rise steeply. It has been often noted that globalization creates a schism between economics and politics. Perhaps it should be added that the breach is also extremely injurious to ethics, for moral considerations are systematically ignored by globalization processes. Under the influence of globalization, national politics are expected to abide by international standards without losing tuition over, and social responsibility for, their subjects. Nations are encouraged to reduce export barriers and favor imported goods, keeping local wages competitively low and discouraging labor demands. The State apparatus should not be onerous, but avoid being too meagre and risk triggering social unrest. Under a mantle of respect for local needs and a show of social sensibility, political weakness and cultural uniformity produce the "one-dimensional man" (Marcuse 1954), submerged in an amorphous society swimming in shapeless, "liquid modernity" (Bauman 2004).

Economic and political globalization has blunted the effectiveness of national policies, at times employing financial pressure to force legislations that are not in the best interest of local governments—privatization of medical services, patent agreements, labor legislation, biomedical and population genomics research policies. In an effort to foster democratic participation in the international forum, public deliberation is suggested *regardless of political citizenship*, hopefully incorporating an extended civil society into "binding laws and then into administrative power." Thus, a *"postwestphalian model of disaggregated sovereignty"* is conceived, leading to "transnational public spheres" administering "transnational public powers" (Fraser 2009, italics in the original). Sober warnings have been issued remarking that broad-based international forums tend to alienate social participation, deriving toward rarefied representations by international officials strong on theory and expertise, but divorced from actual concerns of the people (Gould 2005).

The transnational influence of socioeconomic factors, together with the global tendency to weaken national commitments in the provision of social security and public health coverage, tends to make governmental efforts porous and ineffectual. International aid, were it to honor promises and declarations, would suffice to cover the budget deficiencies of the poor world, but the trend has been to reduce already meager contributions. Scholars have done their share to excuse rich nations from effectively assisting the poor in order not to compromise the well-being of the advantaged. Life-boat logic and the much read article on the "tragedy of the commons" insist that a "finite world can support only a finite population" (Harding 1968). The Malthusian answer is to reduce world population by limiting reproduction and by denying international aid that would reduce the well-being of the affluent and stimulate the poor to multiply.

In these views, redistribution policies of world wealth are not considered: justice is served by reducing destitute populations, not by decreasing consumerism and sharing surplus production. One suggested way of admitting inequities but avoiding global commitments is to modify the status of human rights from humanitarian and universal to political and contingent, or settling for a "minimalist" international negative obligation of avoiding harm, a necessary but evidently insufficient encouragement to reduce global injustice (Daniels 2006). Others have simply given up on the ideas of global justice and health equity, but hold on to a "universal duty to reduce shortfall inequalities in central health capabilities," going on to propose a mixture of national government commitment, institutional efficiency, and individual responsibility (Ruger 2011), or toying with the idea of public–private partnerships to face global health problems. Propositions of this tenor cannot bridge the gulf between philosophy on the one side, and social practices based on *Realpolitik* on the other, nor mend the severed connection between ethnographic knowledge of local cultures and respectful ethically inspired health care policies (Kleinman 1999). Such bridges may not be required by philosophical ethics, but their absence fails to inspire bioethics or any other applied ethics.

Global Ethics

One of the least disputed effects of economic globalization is its influence on politics and culture and on the different understandings of democracy, human rights, and the boundaries of ethical responsibilities. Democracy is so entrenched in Western political thought, that subjecting it to critical deliberation is most unpopular, even though Aristotelian democracy has lost much of its original meaning and flavor ever since "government of the people" has been degraded to a representative format limited to little more than periodical voting exercises. Hard data and trend statistics show that global justice remains a theoretical proposition in the wake of a reality that literally abandons the needy and allows business as usual to deepen inequalities to a point of no return. Globalization explains how nature has become subservient to humans, and how, if the world is the ultimate and sole reality, there are no transcendent values to fall back upon. In sum, globalization depletes humankind of deep beliefs and traditional values, forcing it to develop an unprecedented, purely anthropocentric, and shallow pragmatic ethical views.

An intuitive approach rests on a supposedly current tendency to sympathize with the poor and assist the needy, and some scholars have relied on good will to propose positive attitude toward sharing (Rorty 1996). The better-off are called upon to impartially donate to the needy "until the point at which the marginal value of the next bit of money one might give would do equal good as famine relief and as an increment to one's available spending money" (Singer 2004). According to Singer, once your reasonable requirements are met, you should donate any surplus income to the most needy, wherever they may be.

In the final analysis, global ethics has a very mixed and controversial reception as globalization strays away from ideals such as equity or democratic participation. Citizens are led to believe that the quality of their lives will be increased by global unfettered liberalization of market expansion, stimulating and facilitating consumerism and encouraging profit excesses, as merchandise is produced and massively worn away in a misguided drive to substantially contribute to "human flourishing." Global economic expansion does not neglect to foster local consumerism, that is, globalization requires glocalization, creating an intersection so incongruous that it cannot be grasped in its ethical dimensions. The world needs less ambitious scenarios that elaborate more down-to-earth proposals to eradicate prevalent deprivations.

Inequality and Health

Common sense no less then science has always been aware of the association between poverty, bad health, and reduced life expectancy burdened by high rates of early childhood and premature adult mortality. Poverty and sickness have, in fact, been recognized as loyal bed-fellows since time immemorial, many classical texts depicting how only the rich could afford healthy life styles and enjoy the best available health care.

The influence of socioeconomic conditions on population health has become firmly established to the point where "socioeconomic determinants," now a household word though blunted by academic overuse, conveys the impression that these external factors have gained a naturalistic status resistant to any attempts at change. Conservative thinkers and prosperous populations celebrate social *status quo* as a product of individual effort, furthering the neomalthusian conviction that the poor have brought misery unto themselves as consequence of their indolence, deplorable habits, lack of qualification, and downright refusal to work; consequently, they do not qualify for special assistance (Mead 1986 quoted in Lupton 2005). In this view, proposals to institute universal welfare have "created conditions which made poverty a rational choice" (Murray 1984 quoted in Lupton 2005).

Inequalities in health and health care are not unanimously seen as ethical issues, notably by those estimating that health care is neither a right nor a privilege but a purchasable service (Sade 1971), or that disease is a misfortune but not an injustice (Engelhardt 1979). The opinion prevails, nevertheless, that there can be no social justice without equity in health, as there is no equity in health without social justice (Daniels 2006).

The World Health Organization adheres to the belief that health inequalities—a sanitary problem—become inequities—an ethical issue—when they are avoidable, unnecessary, unfair, and unjust (Whitehead 1992). Leaving aside unfairness as redundant for inequity, the criteria "avoidable and unnecessary" are contingent and subject to interpretations, allowing politicians and economists to determine what is meant by avoidable and to decide in the conference room what can be discarded as unnecessary.

Health inequalities result from a number of complex sets of factors, which need to be considered whenever contemplating remedial intervention: (a) materialists concentrate on resources and external factors; (b) Psycho-social approaches contemplate stress resulting from poor social integration; (c) cultural–behavioral views are concerned with life styles, whereas (d) life-course perspective looks at the temporal aspects of all these factors. Finally, (e) the capability approach has been added but as yet insufficiently analyzed beyond understanding it as a relation between resources available to the individual, and life challenges (Sundmacher et al. 2011).

The question remains open whether it is health that determines how people will fair in managing their life—social selection hypothesis—or, rather, if socioeconomic position conditions health-status and access to health care—social causation hypothesis (Dohrenwend et al. 1992). Dependence may work both ways, but the issue remains confused if it be true that "an individual's chances of life and death… persist even when there is universal access to health services" (Daniels et al. 2006), so that "the differences in access to health care" do "not get us very far" (Peter 2006). Much efforts displayed by academia and international study groups focus on issues concerning global ethics, as well as inequity and its health impact, being unfortunately unable to transcend the think-tank level and having to acknowledge the "dismal nature of extant global health governance," while admitting that "foreign aid often is not aligned with local priorities… Instead, international development assistance for health tends to be framed by donor countries in terms of their geostrategic and philanthropic interests" (Gostin and Mok 2010).

Socioeconomic factors are global in their impact on worldwide injustice, health inequities, and the burdens of disease, but political efforts to foster public health measures and provide medical services can only prosper at the national level. Supporting this view, Britain's Commission on Social Determinants of Health states: "Health, and health equity should become corporate issues for the whole government, placing responsibility for action at the highest level and ensuring its coherent consideration across all policies" (Marmot et al. 2008). Governmental actions are publicly financed by the nation's citizenry, thus confirming the territorial and social unity of State agency. A State-centered policy has to deal with the enormous demands in resources and money required by health care, taking into account that both socioeconomic determinants and public health policies are strongly influenced and constrained by international politics and global economics.

Poor countries lack the resources, or are not willing, to invest in health care or public health programs. Most industrialized countries have been committed to comprehensive national health platforms, but are showing regressive tendencies that limit coverage and introduce exclusion and copayment policies. Medical care and public health measures are often denied to social segments that are very poor; illegal immigrants, the elderly, and other marginalized groups are grossly neglected. The idea of global health equity has been noticeably shelved as major bioethics journals run articles declaring that "access to a sufficient amount of health-related goods is a laudable goal of every State, but not an actual right," for "a right to health-related goods is compatible with the unfortunate likelihood that it will not be honored for the majority of the world's poor for many years to come" (Arras and Fenton 2009).

Public Health as Social Practice

Medicine and public health, according to Henry Sigerist, reflect the cultural conditions of their time. Public health was born as a systematic discipline when disease appeared to be a threat to the strength and productivity of the working class. Political events were the most decisive factors in the development of public health as it came to be understood up to the first half of the twentieth century. In 1648, the Peace of Westphalia created a new territorial order stripping the Holy Empire of its secular power and granting sovereignty to princes who became absolute rulers of the newly recognized nation-states. According to public health historian G. Rosen, the "modern state developed more and more into a centralized national government with a set of political and economic doctrines that in varying degree influenced the administration of public health." Central power developed along the lines of national mercantilism, aimed at subjecting social and economic life to serve the State for the benefit of its rulers.

Culling demographic data was initiated by William Petty as "political arithmetic," and refined by John Graunt to become the discipline of "statistics." Quantified information served to design State policies focused on increasing national productivity and the wealth of sovereigns. A number of political practices were developed, known under the name of *Polizeiwissenchaften*—police sciences—one of the most prominent being the institution of a *medicinische Polizey* developed by Johann Peter Frank (1779). Frank's inspiration is vividly illustrated when, in addressing an audience in Pavia, he quotes France's King Henry IV: "I shall not rest, nor celebrate myself as having been the people's ruler, as long as the peasant is unable to place a hen on his table and enjoy a good meal, in order to restitute his work-weakened strength." Such were the times that prompted Adam Smith to detect "those who consider the blood of the people as nothing in comparison with the revenue of the prince" (quoted in Rothschild 2001).

Medical police inaugurated "the development of liberal political economy [that] was accompanied by an increase, relative to the era of cameralism and mercantilism, in government intervention for the enforcement of health, safety and security" (Carroll 2002). Virchow's dictum "Medicine is a social science and politics is nothing but medicine at a larger scale" set the stage for the birth and development of biopolitics, and to a renewed interest in the intricate relationship between public health and social determinants of health and disease (Foucault 2004).

Many factors were involved in the development of public health as a systematic discipline in the eighteenth century. Plague-ridden Europe, deeply concerned with the ravages of endemic and epidemic diseases, was compelled to develop strict measures of isolation and quarantine. Science promised new insights into mechanisms of transmission and contagion, beginning with Fracastoro's concept of little seeds or *seminaria* as the causal agents of epidemic diseases; knowledge about anatomy and physiology of the human body was rapidly growing. Technological development was influencing mining, manufacturing, and industrial processes, stimulating the early works of Agricola in 1556, and Razazzini in 1770, both describing the hazards of certain occupations and working conditions. Thus, the discipline of occupational medicine was inaugurated but, as yet, failing to envision the social dimensions of disease and medicine that were to flourish from the eighteenth century onward.

As the Industrial Revolution gained momentum, public health developed into a major political force committed to health care in "two major categories: conditions, such as poverty and infirmity, in which the individual has a right to request assistance from the state; and conditions in which the state has the right and the obligation to interfere with the personal liberty of the individual" (Rosen 1993). History teaches that public health was never more than a national enterprise, but also never less.

Historical Demographic and Epidemiological Transitions

Medical practice has to a great extent profited from the fact that wealth and health go together, for many centuries catering to the rich and leaving the poor to be cared for by nonmedical practitioners—barber-surgeons, spicers, apothecaries, and the like. Public health has been equally class conscious, intent on keeping the working class productive and peaceful, hygienists introducing some basic sanitary measures and set out to educate, and force, urban populations to follow healthy habits.

The long-held view that poverty is associated with disease, and that hygiene constitutes a major weapon in combating and preventing epidemic outbreaks and the dissemination of endemic infections—childhood diseases, sexually transmitted conditions, leprosy, tuberculosis, among others—led to the simplistic idea that diseases have a monocausal origin. The doctrine of *causa vera* predominates throughout the nineteenth century, nurturing a notion that once the true cause of a disease becomes known, it will swiftly lead to a specific and highly efficacious therapy—the magic bullet. Although these medical concepts are no longer valid either for infectious or for noncontagious conditions, monocausal thinking persists, for example, in the search for disease-causing genes or in pharmaceutical research for specific therapeutic agents and the development of pharmacogenomics. Monocausal thinking tends to ignore the complex relation between poverty and unhealthy living conditions, high prevalence of disease, and premature death worsened by unequal access to medical care and lack of social security.

Infectious diseases were more prevalent in lower social classes and, as socioeconomic conditions in Europe began to improve in the nineteenth century, diseases such as tuberculosis rapidly declined, long before medicine introduced specific therapy and vaccination (McKeown 1979). The improvement of living conditions in the industrialized world and the demographic transition to lower birth rates and increased life expectancy, together with the epidemiological transition from infections to chronic and age-related diseases, were important factors in turning the focus of public health away from a social toward a clinical activity centered on the individual, giving birth to an epidemiological turn in disease prevention and health promotion.

In most countries of the Western world, the twentieth century witnessed a change in the kind of diseases that affected the population. History has rendered ample tribute to scientific and technological developments that transformed many killer infections into brief, easily dominated sickness episodes, or managed to prevent their occurrence altogether. Bacteriology, virology, immunology, and the discovery or synthesis of potent antibiotic agents have helped control infectious diseases, and contributed to reduce infant mortality and to substantially lengthen life expectancy. As people grow older, they become prey to noninfectious diseases, initially called "man-made" or degenerative morbidity, now more graciously known as chronic diseases. Avoiding the term man-made should not obscure the fact that accidents, violence, and ecological deterioration increasingly cause morbidity and mortality originated in human activities.

Perhaps, "if social and behavioral changes which parallel and propel, the epidemiological transition" are emphasized, one should talk of a health transition, for not only disease prevention but also the modern concept of health is in flux (Beaglehole and Bonita 1999). The implications of these diverse meanings reinforce the suspicion that ethics is becoming an increasingly decisive element in deliberations on public health issues, moving from an academic discipline to becoming socially participative, thus giving birth to a transition in its own right.

Transitions are untidy, and whereas some nations do, in fact, show a predominant prevalence of chronic diseases, others continue to be plagued by malaria, dengue, and other major infections. Most populations show a mixed composition of disease prevalence, and some epidemiologists are referring to a new transition to account for the emergence of hitherto unknown infections, the virulent reemergence of old ones, and the combination of nontransmissible and contagious diseases. These changes have strained the causal thinking of epidemiologists, making them imagine new pathogenetic models based on multicausality and complexity, fueling cognitive uncertainty, and unconvincing preventive and health-promoting policies. As empirical knowledge ramifies and dilutes, public health issues become growingly subject to disputes, unimaginative cost-benefit evaluation, and the disturbing influence of nonsanitary forces, thus creating a fertile ground for much needed qualitative assessment and ethical deliberation.

Perhaps, "if social and behavioral changes which parallel and propel, the epidemiological transition" are emphasized, one should talk of a health transition, for not only disease prevention but also the modern concept of health is in flux (Beaglehole and Bonita 1999). The implications of these diverse meanings reinforce the suspicion that ethics is becoming an increasingly decisive element in deliberations on public health issues, moving from an academic discipline to becoming socially participative, thus giving birth to a transition in its own right.

References

Arras, J. D., & Fenton, E. M. (2009). Bioethics & human rights: Access to health-related goods. *The Hastings Center Report, 39*(5), 27–38.

Bauman, Z. (2004). *Liquid modernity*. Cambridge UK: Polity Press.

Beaglehole, R., & Bonita, R. (1999). *Public health at the crossroads*. Cambridge New York: Cambridge University Press.

Carroll, P. E. (2002). Medical police and the history of public health. *Medical History, 46*(4), 461–494.

Daniels, N. (2006). Equity and population health: Toward a broader bioethics agenda. *The Hastings Center Report, 36*(4), 22–35.

Daniels, N., Kennedy, B., & Kawachi, I. (2006). Health and inequality, or, why justice is good for our health. In S. Anand, F. Peter, & A. Sen (Eds.), *Public health, ethics, and equity* (pp. 63–92). Oxford: Oxford University Press.

Dohrenwend, B. P., Levav, I., et al. (1992). Socioeconomic status and psychiatric disorders: The causation-selection issue. *Science, 255*(5047), 946–952.

Engelhardt, T. E., Jr. (1979). Right to health care: A critical appraisal. *The Journal of Medicine and Philosophy, 4*, 113–117.

Foucault, M. (2004). *Naissance de la biopolitique*. Paris: Seuil/Gallimard.

Fraser, N. (2009). *Scales of justice*. New York, Chichester: Columbia University Press.

Gostin, L. O., & Mok, E. A. (2010). Innovative solutions to closing the health gap between rich and poor: A special symposium on global health governance. *The Journal of Law, Medicine & Ethics, 38*(3), 451–458.

Gould, C. C. (2005). Beyond minimalism in human rights and democracy: A response to Nickel and Bohman. *Journal of Global Ethics, 1*, 224–238.

Harding, G. (1968). The tragedy of the commons. *Science, 162*, 1243–1262.

Kleinman, A. (1999). Moral experience and ethical reflection: Can ethnography reconcile them? A quandary for "The New Bioehics". *Daedalus, 128*, 69–96.

Lupton, D. (2005). *Health, civilization and the State*. London: Taylor & Francis.

Marcuse, H. (1954). *One-dimensional man*. Boston: Beacon Press.

Marmot, M., Friel, S., et al. (2008). Closing the gap in a generation: Health equity through action on the social determinants of health. *Lancet, 372*(9650), 1661–1669.

McKeown, T. (1979). *The role of medicine*. Oxford: Basil Blackwell.

Milanovic, B. (2008). Developing countries worse off than once thought. *YaleGlobal* (February 11), 1–2.

Miller, D. (1999). *Principles of social justice*. Cambridge, London: Harvard University Press.

Peter, F. (2006). Health equity and social justice. In S. Anand, F. Peter, & A. Sen (Eds.), *Public health, ethics, and equity* (pp. 93–106). Oxford, New York: Oxford University Press.

Rorty, R. (1996). Who are we? Moral universalism and economic triage. *Diogenes, 44*(173), 5–15.

Rosen, G. (1993). *A history of public health*. Baltimore: The Johns Hopkins University Press.

Rothschild, E. (2001). *Economic sentiments*. Cambridge, London: Harvard University Press.

Ruger, J. P. (2011). Shared health governance. *The American Journal of Bioethics, 11*(7), 32–45.

Sade, R. M. (1971). Medical care as a right: A refutation. *The New England Journal of Medicine, 285*(23), 1288–1292.

Singer, P. (2004). Outsiders: Our obligation to those beyond our borders. In D. K. Chatterjee (Ed.), *The ethics of assistance* (pp. 11–32). Cambridge, New York: Cambridge University Press.

Sundmacher, L., Scheller-Kreinsen, D., et al. (2011). The wider determinants of inequalities in health: A decomposition analysis. *International Journal of Equity Health, 10*(1), 30.

Taylor, C. (1994). *Justice after virtue*. Notre Dame: Notre Dame Press.

Weinstock, D. M. (2006). The real world of (global) democracy. *Journal of Social Philosophy, 37*, 6–20.

Whitehead, M. (1992). The concepts and principles of equity and health. *International Journal of Health Services, 22*(3), 429–445.

Chapter 2
Rights and Duties, Needs, and Merits

Abstract International Declarations and Covenants on human rights pretend to have universal reach, at the same time acknowledging that basic rights can only be cultivated and protected at the national level. This universality has been questioned by non-Western cultures, and by those pointing out that marginalized social groups, by being deprived of citizenship, are subject to violation of their freedom-based negative human rights, and unable to claim positive rights—basic goods. Furthermore, poor- and middle-income countries may lack the political will to protect basic rights, or be unable to muster resources to provide essential goods and services, thus leaving essential needs unattended.

Distributive justice searches for criteria to reduce inequalities and create fair social conditions. Liberal politics defend justice based on merit and access to equal opportunities as the most appropriate criteria for allocating scarce resources. Since poor populations lack the capabilities to behave meritoriously and achieve socially respectable positions, merit cannot lead to more justice. Addressing essential needs is the objective and foremost criterion to allocate resources appropriately, and ought to be applied wherever dire needs are unattended.

Keywords Basic rights • Distributive justice • Equal opportunities • Merit • Need • Risk

Justice Based on Human Rights

Human rights have been with us since the French Revolution, acquiring universal status initiated by the United Nation's Universal Declaration of Human Rights [1948], later buttressed by the International Covenant on Civil and Political Rights, and the simultaneously presented International Covenant on Economic, Social and Cultural Rights [1976] (Gostin and Archer 2007). Human rights have been proclaimed and defended at global, regional, national, and local levels, and presented in the form of declarations, conventions, treaties, recommendations, and guidelines.

M. Kottow, *From Justice to Protection: A Proposal for Public Health Bioethics*,
SpringerBriefs in Public Health 1, DOI 10.1007/978-1-4614-2026-2_2,
© Miguel Kottow 2012

Cosmopolitan views have their greatest appeal at the humanitarian level as they navigate theoretical territories, but they lose plausibility and conviction when the appealing but weak plea for rights obscures the more effective and binding claim that correlative obligations ought to effectively secure the demands of right holders (O'Neill 2004). Global justice defenders are notoriously vague on this issue, mostly falling back on the State as the agent responsible for implementing human rights and just procedures at the national level, but eluding the problems and contradictions involved in the brute fact that globalized economics are major factors that make States weak and porous, thus contributing to the persistence of international inequality (Beck 1998).

There seems to be a general, though tacit, agreement that the national level is in the best position to enforce human rights and to sanction violations within its territory. And yet, "many countries have not incorporated human rights provisions or norms into their national laws and politics" (Gable 2007). As a consequence "democracies routinely contain millions of disenfranchised people, namely children and those who are for a variety of other reasons judged to be incompetent" (Weinstock 2006). Some nations emphasize negative rights, but barely mention substantive rights like education or health, while others present comprehensive lists of economic, social, and cultural rights, too often honoring them in the breach. All Latin American countries include a right to health care or health protection (Fuenzalida-Puelma and Connor 1989), but national health services are not up to the task of meeting such commitments, once again illustrating how rights may become empty claims if not correlated with obligations of fulfillment.

The universality of human rights has been questioned by non-Western cultures who resent their marked individualism and indifference to the multiple commitments of human beings in a cosmos of social and natural bonds where a person's life-span is but an episode in transcendent holistic processes. A number of philosophers have pointed out that rights are not universal, seeing that they can only be claimed from the status of recognized citizenship (Rancière 1998). Children, the mentally impaired, the poor and marginalized, and illegal immigrants are all groups that lack the empowerment to voice their claims for recognition as legitimate right holders. The advocacy of rights for the distant or the absent—future generations—remains contested and unheeded. To be consistent, deep ecologists claim, global ethics must go beyond the idea of human rights, incorporating the need to protect nonhuman living beings, nature itself, even macrocosmic Gaia.

Rights necessarily correlate with the obligation to acknowledge, respect, and fulfill them. A right that is not claimed remains virtual and ineffective, whereas claiming a right presupposes an addressee whose unfailing duty is to satisfy the petitioner (O'Neill 1998). Dissident voices maintain, to the contrary, that basic rights do not "necessarily impose second- or third-party positive duties of protection or provision," which could well be honored by "voluntary institutions" (Cohen 2004). Allowing voluntariness to determine the fate of human rights is tantamount to denying their relevance in *Realpolitik* and their lack of influence on social practices, as they are displaced into the discretionary arena of political and social reality: "If the ideal conception of human rights must be embedded in an institutional conception in order for human rights to become truly effective and action guiding…[subject] to degrees of scarcity and…particular cultures and states…[then] they do not have universal reach. In other words, institutional human rights are not, strictly speaking, unmodified *human* rights. They will, rather, bear much more resemblance to *political* rights, which are recognized by particular states on the basis of their own particular political culture and value priorities" (Arras and Fenton 2009).

Facing Social Justice

Philosophers may delve in moral quandaries, but applied ethics must come up with at least tentative answers. Justice-based responses are found wanting, all the more so if equality is dismissed for having "no inherent or underived moral value at all" (Frankfurt 1997). When justice is neglected as an inappropriate frame of reference to approach the perplexities inherent in distributing scarce resources, applied ethics must search for a set of criteria that hopefully will allow apportioning in a morally legitimate way. Social justice advocates fair criteria for the distribution of scarce resources to legitimate holders of positive rights, such as education, health care, medical services, and social security. Many criteria have been proposed: merit, need, first come first served, social worth, age, expected utility, or lottery. Equitable distribution has inspired some of these criteria, but health-related goods and services should be less concerned with justice than with ranking access and availability in accordance with the importance of presenting problems. Only merit and need will be discussed, the first because it is the preferred criterion in developed societies, whereas need, it will be argued, is the only ethically valid one in societies or population segments that live in deprivation.

Merit and Luck Egalitarianism

"Justice is a disposition to give to each person, including oneself, what that person deserves and to treat no one in a way incompatible with their desserts" (MacIntre 1984; MacIntyre 1988). This description presents a number of difficulties, the most obvious one being that desert is seen as equivalent to merit, although they are not analogous. Desert applies in situations "in which someone is responsible for the results he or she brings about," whereas merit refers "broadly to a person's admirable qualities" (Miller 1994). Much more troubling is the idea that just distribution is to be assigned according to individual comportment or qualities, a proposal known as justice in inequality. Meritocracy shares with justice in inequality the conviction that desert/merit should operate *after* the basic needs have been met by social institutions treating all citizens as equal. Once opportunities are equally accessible, individuals should fare according to their freely exercised ability and efforts to obtain social privileges and nonessential material goods. Minimum decent coverage is, unfortunately, disregarded when economically troubled liberal governments or staunch libertarians gain the day.

At least three objections come to mind when evaluating the liberal celebration of meritocracy: First, basic needs are glossed over and often dismissed as lacking objectivity and firm intercultural validity; second, social institutions do not guaranty basic equality nor equal opportunities, quite to the contrary, they are steeped in unavoidable power differentials and asymmetrical relations (Young 2011). Being organized around work and production, societies unavoidably develop hierarchies and income gradients (Rancière 1998). Third, beyond a minimum of equality, individuals need to be supportively empowered to freely employ their capabilities in securing their place in society and the pursuance of their life project (Sen 2000; Nussbaum 2006).

Extreme libertarians may fail to accept that merit can only become a criterion of fairness, once basic needs and empowerment to seize equal opportunities have been secured. The fact that pockets of poverty prevail in rich countries proves that merit schemes are unfair at the basic level where they simply do not operate. But meritocracy does have its virtue, for it caters to the middle classes who are chronically dissatisfied about their duties and the fact that their tax burdens only buys them basic social security. If honoring merit will spur drives toward excellence and prestige, it may be a welcome addition to need-based equity.

Meritocracy blends with the general belief that people are responsible for their life, their place in society, and their relative well-being as compared to others. In the same vein, the worse-off will have no one to blame but their own inferior performance and lack of enterprise to seize opportunities. Even the most responsibility-centered advocates will admit that external circumstances, which in these theories go under the name of luck, strongly influence the range of choices an individual commands. Human beings are not born equal, genetics, moreover, audaciously asserting that molecular determination underlies not only corporeal anatomic and functional constitutional features, but also preferences and conducts. "Brute luck" points at the circumstances beyond personal control, which limit the exercise of autonomy, as distinct from "option luck" which describes the vagaries of results brought about by each person's choices. Unlucky circumstances for which a person cannot be held responsible ought to be compensated in a fair society.

Luck egalitarianism, also known as equality of fortune, is "a hybrid of capitalism and the welfare state" (Anderson 1999), suggesting that unfortunate circumstances, being undeserved, ought to be removed or mitigated in order to secure a just allocation of essential material and functional resources, as well as empowerment; in addition, people ought to command a basic set of capabilities to compete in the market for their welfare, security, and wealth. The dividing line between undeserved misfortune and self-responsible outcomes is arbitrary and impossible to agree upon, for "structured social positions" and circumstances plays a major part not only at the starting-gate, but also throughout the whole length of peoples' lives (Young 2011). To equalize the impact of uneven natural endowment, individuals should command a set of capabilities allowing them to function as human beings, as productive members of society, and as participating citizens in a democratic state. Whatever each person actually does with these capabilities is a matter of free choice and therefore her own responsibility; unfortunate choices are not to be compensated unless they compromise the basic capabilities that are inalienable (Anderson 1999).

By flatly stating that democratic equality matches "the remedy to the injustice: if the injustice is exclusion, the remedy is inclusion," an important bridge is spanned toward the ethics of respect and recognition, and subsequently to an ethics of protection. Respect trumps over equality, for every "person should be accorded the rights, the respect, the consideration, and the concern to which he is entitled by virtue of what he is and of what he has done." Respect acknowledges individuality, which equality fails to do because it addresses disparity instead of actual need (Frankfurt 1997). In order to respect someone, three levels of recognition must be obtained and be correlated with corresponding obligations of fulfillment: a person's "needs and desires," her status as a moral agent, and the capabilities of social integration and participation. Individual needs and desires are to be met with love and care, moral status is to be respected, while social agency demand "solidarity" or "loyalty" (Honneth 1997). Although the ethics of recognition does consider basic needs, it cannot account for the moral quandaries of the distant, the absent, and the silent, who will not be recognized, for they are voiceless. Without saying so, it allows the liberal idea of earning respect without considering that earning something requires a set of basic capabilities.

In an elegant turnabout, A. Margalit believes that the values of equality and freedom depend on decent societies and institutions that do not humiliate their constituents (Margalit 1997). People are humiliated if they are mistreated, their merits are not recognized and, above all, their needs are ignored. The ethics of recognition has a strong theoretical appeal, but its translation into practice remains unexplored.

Need

Need is an often posed measure for just distribution, but objections are recurrently presented to hinder necessity from becoming a full-fledged and undisputed criterion for an ethics intent on assuaging and relieving deprivation and suffering caused by lack of essential goods. Needs are an unwanted but inevitable consequence of resources being finite and insufficient for universal satisfaction. Therefore, availability of essential resources would become a litmus test of a fair society that acknowledges and attends basic needs. Health care needs are seen in the humanitarian view as a "disturbance in health and wellbeing," as opposed to the—"realistic view" that recognizes as need only those conditions ably met by medical intervention that "alters the prognosis of the disease in a favorable way at reasonable cost" (Acheson 1978). Both views are flawed from the ethical perspective: disturbance is too vague a term to vindicate a duty or a right, and conditioning need to the ability of satisfying it is inconsistent: did suffering from AIDS only qualify as need once antiviral agents became available? At the most, it could be said that the obligation to satisfy a need cannot obtain unless an effective intervention exists—can implies ought—but the actual suffering of the needy remains undaunted whether relief exists or not. It is further objected that needs lack objectivity and are a matter of personal preferences and idiosyncrasies. "Which claims count as needs in the first place; which needs give rise to demands of justice; and how to establish priorities among different qualifying claims: these questions raise complex ethical issues that belie the apparent simplicity of "to each according to his needs"" (Miller 1999). These objections miss the moral point that certain needs are basic and universal, and that resources probably would if justly distributed.

Bodily needs are no doubt essential. Food and water are indispensible elements for survival, and if the autopoietic [self-generating] vital processes fail to perform—as in disease—the living system succumbs unless external assistance—therapeutic care— is instituted. The basic needs of living beings are universal and therefore public, that is, generally accepted. These needs are not negotiable, neither amenable to lexical ordering nor postponement, for they cannot be displaced from their principal position by other needs. Being universal, bodily needs are context independent and distance neutral, equally valid if directly recognized or reported from remote regions.

A reasonable and straightforward understanding of needs defines them as requirements of an organism in order to live a healthy life, but such a definition will not do since it relies on "healthy life" as a criterion that is itself undefined and controversial. Other descriptions are more ambitious, like Maslow's hierarchy of psychological needs or preferences. Broadly relating needs to undefined and unspecified social integration make it all the more difficult to identify those necessities that are primary, inherent, or fundamental. Nor is much gained by resorting to Marx's vivid aphorism "to each according to his needs" or his definition of humans as "creatures of need." Comprehensive definitions bring forth subjective elements that invite polemics and foreclose agreements as to what ought to qualify as basic need.

Occasional approaches have set priorities in a realistic bottom-up perspective. Thus, the Framework Convention on Global Health (FRGH) hosted by the O'Neill Institute for National and Global Health Law proposes to "set priorities so that international assistance is appropriately directed at meeting basic survival needs" and directing resources "so that all elements of the health sector can perform their core functions and meet the population's basic needs in a sustainable manner" (Gostin and Mok 2010). Down-to-earth recognition of unmet essential needs is a more promising approach than proclaiming officially favored unrealistic goals like global justice or universal well-being, hopefully meeting approval and inspiring action.

Essential human necessities go beyond the stark fulfillment of corporeal requirements since humans must relate socially and integrate to the milieu they live in, developing the empowerment to autonomously exercise certain capabilities in pursuance of existential projects. Lack of basic social capabilities is also a situation of dire need, for the disempowered will not survive in a structured and complex society.

Intent on the provision of bodily survival needs—food, shelter, medical care, basic education—and securing social agency that safeguards the luckless from irremediably precipitating into primeval impotence, societies must institute a safety net in form of unemployment subsidy, emergency loans, disease treatment, rehabilitation, old-age support. Even such a coarse mapping of needs as bodily essential and existentially basic should help design a social agenda that recognizes obligations to the needy—those in dire bodily need—and those requiring empowerment to maintain autonomous functioning in pursuance of a life-project. Though captive in the language of justice, a similar proposal is echoed in the statement that "the foundation of deontology is *the desire to live well with and for others in just institutions*" (Ricoeur 1992, italics in the original).

Need must be related to the needy, an apparent tautology that stresses the objective quality of basic needs, and should discourage those who condemn need-talk to triteness by claiming that arbitrary idiosyncrasies posing as necessities are not to be socially respected. Certainly, even basic needs may to some degree be contextual: in poor communities survival is the primordial need, whereas in more developed societies it is essential to be capable of mastering certain basic skills in order to gain support and social integration. Philosopher Agnes Heller has theorized on necessity, attending to J.P. Sartre's distinction between *manqué*—deficiency that reflects need—and *project*—life plans referred to satisfaction. She has also made the point that political and cultural systems tend to foist needs on people in a way that contradicts the most elementary necessity of being one's own master (Heller 1993).

Needs are to be understood from a bottom-up perspective that recognizes their stark reality and frees the issue from as much contextual ballast as possible to gain them universal acceptance as basic to survival. Grass-root bottom-up health care planning as practiced in Oregon is a case in point, as is the report about deliberating low-income individuals that have been known to voice their hope of directing "planning and budgeting services to most effectively address the immediate needs of the community" (Pesce et al. 2011). It might be argued that context-free conceptual definitions of need are contradictory in relation to tangible basic goods and services, but a robust ethical concept of need ought to resist the bloodless vagaries of financial narrowness and political moods.

One the most discussed forms for allocating scarce medical resources is the strategy of triage, inherited from the Napoleonic wars. Essentially, triage suggests reserving scarce resources to those who will actually be rescued if attended to. In war time, this meant that the severely injured were left behind because they could not be salvaged, while the lightly wounded were also neglected, since they would probably heal without intervention. If looked at through the optics of need, triage represents a need-based criterion of allocations, since resources are concentrated on those who require them most, for their future is the one that directly depends on receiving care.

Poverty

Looking closer at poverty reveals some unsettling realities. A country may be at the top of the list in terms of *per capita* income and overall human development—a concept used by the United Nations Development Program—yet perform badly in poverty rating or population health ranking, thus illustrating that equality is foremost a matter of distribution rather than aggregate wealth. Equally baffling is the fact that minimum wages may be insufficient to keep families out of absolute poverty, in which case they are, in fact, subminimal (Williamson and Reutter 1999).

Basic needs are related to biological deprivation, whereas comparative poverty, although including needs, is associated with social indicators that measure social inclusion/exclusion (Marlier and Atkinson 2010). "Poverty is not an absolute state"; individuals and families whose resources over time fall seriously below the average of members of the community in which they live "are in poverty" (Townsend 2010).

Economists distinguish between absolute poverty related to material want, and relative poverty referred to social deprivation (Williamson and Reutter 1999). The absolute poor are deprived of the essential goods necessary to cover their biological needs, while the relatively poor are insufficiently empowered and devoid of opportunities on average available to other members of their society. Health care programs should address need and deficient empowerment inasmuch as both kinds of poverty are detrimental to health. Resources being scarce, public health must first attend those in dire need—the undernourished, those weakened by disease, those befallen by catastrophic illness, the unprotected—before attending the health problems of social deprivation—emotional distress, mental illness—which may have to be postponed for lack of funds, but cannot be ignored.

Primary Goods

Arguments from insufficient resources are specious as long as short supply of basic goods is caused by grossly unequal distribution and not by absolute scarcity. The world's wealth increases, and so does the population of the needy. Discrimination, exclusion, illegal immigration, and economic crisis are some factors that perpetuate pockets of destitute populations even in the most affluent nations. Superfluity and consumerism flourish side-by-side with people in dire needs as shown by the fact that humanity produces enough food to provide over 3,000 calories per day to every human being on earth. This massive food production is consumed, abused, or wasted in such a way that one third of the human population is undernourished, half of which receives less than the minimum food uptake necessary for body maintenance (Nadakavukaren 1984).

The enormously complex issue of relating needs, primary goods, empowerment, and citizenship, has tasked the efforts of influential philosophers. Rawls elaborates the specification of citizens' needs as "a construct worked out from within a political conception and not from within a comprehensive doctrine." His list of primary goods is a "political understanding of what is publicly recognized as citizens' needs," and yet, one of its headings is "income and wealth," hardly a basic need (Rawls 1996). Rawls is dealing with members of society who enjoy the status of citizens, and rely on a functioning democracy that facilitates agreements on an equal standing—the original position: each participant enjoys equal rational expediency to engage in reflective equilibrium and reach overlapping consensus. This is, of course, a very selective view of humanity. A. Sen is less interested in listing and indexing primary goods, than in describing their functional value to reach empowerment and develop capabilities, but some primary goods are so essential that they are transcendental, that is, necessary conditions that must initially obtain before an individual can decide what goods he wishes to engage in pursuing his choice of achievements. The capabilities freely employed are rooted in the basic empowerment every person requires to pursue whatever life-plan is freely chosen.

References

Acheson, R. M. (1978). The definition and identification of need for health care. *Journal of Epidemiology and Community Health, 32*(1), 10–15.

Anderson, E. (1999). What is the point of equality. *Ethics, 109,* 287–336.

Arras, J. D., & Fenton, E. M. (2009). Bioethics & human rights: Access to health-related goods. *The Hastings Center Report, 39*(5), 27–38.

Beck, U. (1998). *Was ist Globalisierung?* Suhrkamp: Frankfurt a.M.

Cohen, A. I. (2004). Must rights impose enforceable positive duties. *Journal of Social Philosophy, 35,* 264–276.

Frankfurt, H. (1997). Equality and respect. *Social Research, 64,* 3–15.

Fuenzalida-Puelma, H. L., & Connor, S. S. (1989). *The right to health in the Americas.* Washington: Panamerican Health Organization.

Gable, L. (2007). The proliferation of human rights in global health governance. *The Journal of Law, Medicine & Ethics, 35*(4), 534–544. 511.

Gostin, L. O., & Archer, R. (2007). The duty of States to assist other States in need: Ethics, human rights, and international law. *The Journal of Law, Medicine & Ethics, 35*(4), 526–533. 511.

Gostin, L. O., & Mok, E. A. (2010). Innovative solutions to closing the health gap between rich and poor: A special symposium on global health governance. *The Journal of Law, Medicine & Ethics, 38*(3), 451–458.

Heller, A. (1993). A theory of needs revisited. *Thesis Eleven, 35,* 18–35.

Honneth, A. (1997). Recognition and moral obligation. *Social Research, 64,* 16–35.

MacIntre, A. (1984). *After virtue.* Notre Dame: University of Notre Dame.

MacIntyre, A. (1988). *Whose justice? Which rationality?* London: Duckworth.

Margalit, A. (1997). Decent equality and freedom: A postscript. *Social Research, 64,* 147–160.

Marlier, E., & Atkinson, A. B. (2010). Indicators of poverty and social exclusion in a global context. *Journal of Policy Analysis and Management, 29,* 285–304.

Miller, D. (1994). *Virtues, practices, and justice.* Cambridge, London: Harvard University Press.

Miller, D. (1999). *Principles of social justice.* Cambridge, London: Harvard University Press.

Nadakavukaren, A. (1984). *Our global environment.* Longrove: Harvard University Press.

Nussbaum, M. C. (2006). *Frontiers of justice.* Cambridge, London: TheBelknap Press of Harvard University Press.

O'Neill, O. (1998). *Towards justice and virtue.* Cambridge: Cambridge University Press.

O'Neill, O. (2004). *Autonomy and trust in bioethics.* Cambridge New York: Cambridge University Press.

Pesce, J. E., Kpaduwa, C. S., & Danis, M. (2011). Deliberation to enhance awareness of and prioritize socioeconomic interventions for health. *Social Science & Medicine, 72*(5), 789–797.

Rancière, J. (1998). *Aux bords du politique.* Paris: La Fabrique-Éditions.

Rawls, J. (1996). *Political liberalism.* New York: Columbia University Press.

Ricoeur, P. (1992). *Oneself as another.* Chicago, London: The University of Chicago Press.

Sen, A. (2000). *Development as freedom.* New York: Alfred A. Knopf.

Townsend, P. (2010). The meaning of poverty. *The British Journal of Sociology, 61*(Suppl. s1), 85–102.

Weinstock, D. M. (2006). The real world of (global) democracy. *Journal of Social Philosophy, 37,* 6–20.

Williamson, D. L., & Reutter, L. (1999). Defining and measuring poverty: Implications for the health of Canadians. *Health Promotion International, 4,* 355–364.

Young, I. M. (2011). *Responsibility for justice.* Oxford, New York: Oxford University Press.

Chapter 3
Public Health Transitions and the "New Public Health"

Abstract Beginning in the 1940–1950s, public health has suffered a major transition to a New Public Health focused on acknowledging socioeconomic risks to health, at the same time internalizing them to the individual who now is seen as a risk carrier. Population-based prevention became Preventive Medicine incorporated in the clinical encounter, developing diagnostic tests and explorations to detect risk-laden predispositions, preclinical conditions, even suspicious genetic markers. Individuals are medicalized into "healthy patients," subjected to routine control check-ups, preventive medication, indications to lead healthy life style and show correct comportment. Public health prevention becomes an individual responsibility aimed at avoiding risk and preventing disease. Naturally, the well-off can adapt their life styles and incur in the extra expense of having routine check-ups, including high-tech procedures, going to spas and gym workouts, and buying expensive pharmacological stabilizers. Furthermore, if risk is internalized, there is no major pressure to publicly intervene in social and environmental health-threatening conditions. Accordingly, international organizations have given up on major ecological interventions, now proclaiming "adaptation and mitigation" policies which are not far from a conservative business as usual attitude.

Deprived of resources, public health loses its traditional agenda of collective prevention against diseases, having to witness the lack of governmental policies on security and protection of the less affluent, who have become insolvent and unable to obtain social support.

Keywords Medicalization • New Public Health • Prevention • Risk internalization • Self-responsible health care

M. Kottow, *From Justice to Protection: A Proposal for Public Health Bioethics*,
SpringerBriefs in Public Health 1, DOI 10.1007/978-1-4614-2026-2_3,
© Miguel Kottow 2012

The three historical transitions in public health originated in major demographic and epidemiological changes. The ravaging epidemics that plagued Europe up to the nineteenth century, kindled by "military explorations, trade, and travel," set the stage for "exposing nonimmune populations to new diseases" (Fee 1993). Plagues finally waned as nation-states systematically instituted hygienic and sanitary measures in an effort to curb the pathogenic effects of urban crowding, poverty, and exploitative labor conditions that might fuel social unrest. The second transition, from infested populations to individuals affected by infectious diseases, led to observation of patients in hospital-care based on positivistic medicine and the development of bacteriology which, in turn, triggered public health measures based on scientific knowledge and clinical experience. The third transition, from infectious to noncommunicable diseases occurred in the wake of industrialization and improved living conditions, as illustrated by the steep decline of tuberculosis long before specific immunization and antibiotic therapy came into being (McKeown 1979).

The Transition from Public to Private Health Care

The public health transition actually in course originates in the predominance of liberal market-centered economy reigning without competition, and the conversion of the State from entrepreneur to regulator emphasizing social stability as the basis for individual self-care. Salient contemporaneous issues are predominantly disputed in terms of values: is environmental deterioration morally sustainable? Does progress lead to more or to less justice? Is industrial development a universally desirable goal? Is public health a promoter of health or a *status quo* stabilizer? Has medicine become another market-centered innovator redefining health care and medical intervention as a commodity rather than a service?

The French and American eighteenth century revolutions had recognized "health citizenship" as a right, but also as an obligation. The individualistic trend of modernity places an ever increasing load on individual citizens to take responsibility for their security, health, and material wants. Contemporary individualism takes an economy-centered view, recognizing the citizen's rights to be an autonomous consumer who accepts the responsibility and financial burden of remaining healthy. Just as medical police imposed measures to keep workers robust and productive, present day liberalism commits individuals to care for their health and be active consumers who energize the economic system.

The prevalent tendency is to replace strong governmental commitment to public health, with market-oriented private medical enterprise. Healthy individuals are registered on sick lists, for the detection of predisposing features recruits them as lifelong paying consumers of health-related products. Medicine and the public are united in health and in sickness until death do them part. Government loses its grip on social security, and health care becomes more expensive and less subsidized, the citizenry of most countries being faced with the necessity of providing their own security and financing the educational and medical needs of their family.

Neoliberalism has developed a new form of governance that displaces social policy "from the constitution of a right to the management of a lifestyle." Risks detected as external factors are shifted to become individually endogenous predispositions, while social protection is replaced by "moralization" of human behavior in all wakes of life, notably work and health. "With 'social reconstruction' one enters into an era of monitoring of the behavior and lifestyles of those receiving medical care, who are obliged to take responsibility for appropriate changes in their own conduct" (Lazzarato 2009).

When M. Thatcher pronounced her infamous verdict: "There is no such thing as society, there only are individuals and family," she was voicing the views of much respected economists like F. Hayek and members of the Chicago school. Deeply suspicious of socialism, conservative thinkers like Popper and Oakshott helped stage a blend of politics that seemed to justify the State's inclination to share the economic burdens of social security with the private sector. In Britain, homes and care centers of the elderly, children, and the mentally ill were privatized to the extent that the share of personal services that Government passed on to for-profit enterprises rose from 11% to 34% in the years 1979–1996 (Judt 2010).

Two major cultural trends contributed to the ready acceptance of an individual, private-based health care adoption of the traditional public health agenda of disease prevention and health promotion. The human body became an object of cult and care, allowing a number of perspectives to emerge emphasizing a diversity of corporeal aspects: gender, sexuality, cleanliness, discipline, recreation, and death (Lupton 2003); as well as categories such as the lived body, the disabled body, the enhanced body, and others could be added. Secondly, nineteenth century's Hospital Medicine has been superseded by Surveillance Medicine that includes observation and exploration of a seemingly healthy population, eroding the distinction between the normal and the ill, identifying external risk factors but placing them within the individual "predisposed" body, and recreating an ontological view of disease that reveals the influence of genetics, social factors, personal behavior, and life style (Armstrong 1995). These are the elements of the most recent public health transition known as Preventive Medicine, "preventivist paradigm," or New Public Health.

The "New Public Health"

During the second third of the twentieth century, the United States' medical establishment observed with mounting uneasiness how Central European countries instituted National Health Services, and socialist countries developed statist medical care. The swift response to budding socialist overtures such as the New Deal (1933–1938) was to reinforce private medical care by teaching physicians to embark in preventive measures that traditionally had been the responsibility of public health policies. Clinicians were trained to develop "Preventive Medicine" as part of their medical practice, urging people to submit to periodical check-ups and exploratory routine laboratory tests, thus becoming "healthy" patients expected to consult with regularity, take preventive medication, and submit to pharmacological neutralizations and surgical extirpation of eventually disease predisposing lesions. The driving idea of this new approach to clinical prevention has been to engage active individual participation in personal health care by promoting self-care and self-responsibility (Arouca 2003).

In 1974, the Canadian Ministry of Health and Welfare presented the four-tiered "Health Field Concept," based on human biology, environment, lifestyle, and health care organization. Reaffirming the will to provide universal medical and hospital care, and insisting on the need to reinforce "care versus cure," the Canadian report stated: "In addition to the health care system and the people collectively, individual *blame* must be accepted by many for the deleterious effect on health of their respective lifestyles. Sedentary living, smoking, over-eating, driving while impaired by alcohol, drug abuse and failure to wear seat-belts are among the many contributory factors to physical or mental illness for which the *individual must accept* some responsibility and for which he *should seek correction*" (Lalonde 1981). Emphasis has been added to the original text, to highlight the accent it places on life style in health and disease management: the moral category of blame, as well as the consecration of self-responsibility and self-care, become an explicit part of public health's discourse redefining citizens' dwindling health-related rights and growing responsibilities (Petersen and Lupton 2000). A recent follow-up report on health in British Columbia tardily acknowledges accumulating evidence pointing to the limited success in preventing chronic disease by approaches aimed at modifying individual lifestyles and behavioral risk factors (Authority 2011).

Internalization of Risk Factors

Epidemiological research has been diligent in culling data that show associations between the increased incidence of major diseases, and prevalent contextual factors that are supposed to be causally related to morbidity and mortality, investigations being based on studying populations and the deleterious environmental influences they are exposed to. The most convincing association came from the solid statistical proof that smoking significantly raises the incidence of respiratory tract diseases, notably lung cancer. Other associations were suggestive but less direct, such as obesity and sedentary habits predisposing to cardiovascular disorders. Contextual risk factors are unavoidable consequences of modern civilization leading to urbanization, environmental deterioration, and technological determination of social processes and individual life courses. Recognizing "the importance of climate change as a threat to global health" the WHO calls "on the health community to protect health from climate change…by strengthening the resilience of human systems" and driving "the process of setting an agenda for 'health adaptation' to climate change" (Wiley 2010). Climate change is an ongoing process, at best mitigated, constituting a risk factor to which the human body must adapt.

In the 1970s, these "conditions of modern life" were redefined as "unhealthy life styles," meaning that, if external factors were not modifiable, an exercise in introspection would reveal that individuals could detect how susceptible they were to these circumstances and how they could avoid their deleterious effects on health. Doctors began to excel at diligently ordering exhaustive lab tests and actively promoting supportive medication, diets, bodily exercises, and frequent control visits. Risk factors moved from the external world to the personal realm, so that disease prevention was to be managed by individual autonomy, thus constructing a moral duty to be well: "While the 'old' public health strategies focused almost entirely on issues of public hygiene—the cleanliness of the streets, the regulation of industry, sanitation and water supply—the new public health has directed its attentions towards the conduct and appearance of the individual body" (Petersen and Lupton 2000).

Internalization of risk factors and self-prevention seem to disregard the "Prevention paradox": "A preventive measure which brings much benefit to the population offers little to each participating individual" (Rose 1985). Applying population data to individuals is most uncertain and has been often dismissed as grossly inaccurate, all the more so if the individuals targeted do not belong to the studied population, therefore putting to question the external validity of extrapolations.

Internalized risk factors must be detected through clinical, laboratory, and genetic exploration, in search of early changes presumed to indicate a predisposition to eventual disease. An apparently intermediate position has gained the favor of such influential institutions as the US Institute of Medicine of the National Academies, forging public health for the twenty-first century based on three premises: an improved public health infrastructure, programs based on scientifically proven determinants of population health, and a wide private sector engagement in health promotion (Gostin and Bloche 2003). This is a contentious issue containing a vast number of topics that branch out into ethics, politics, and economics; whereas some have considered that economics per se is the unfailing agenda of public health, others argue that "the interdependence of ecosystems and health systems" requiring "reciprocal maintenance" hails a holistic view that reclaims for itself the label of being a new public health (Vandenbroucke 1994).

Knowledge about the interaction between constitutional and environmental factors which might lead to disease is very sketchy, and anticipating whether full-fledged clinical manifestations will at some point evolve from a detected predisposition is little more than guesswork. The notion of health becomes indeterminate, and people living a normal life will be labeled as patients on very flimsy evidence. Objectively detected external risk factors are variably lived as "at risk" experiences. The incommensurability between epidemiologic risk detection and laypeople's perception of risk accounts for the difficulties encountered when medicine explores personal predispositions and encourages compliance to behave like potential patients.

Second thoughts dwell on the ethical solidity of promoting life style adjustments that are only affordable and practicable by the more affluent segments of society. In an effort to reestablish cooperation between self-care and public health, individuals have to be seen as more than singular entities, stressing their citizen status including the rights and duties entailed. Citizens are expected to integrate into their social web, actively contribute to the common weal and, above all, be productive. Devoid of State intervention, self-control is of primer importance to reduce dependency and avoid burdening the fiscal purse, but a debilitated social network will strand the weaker members of society who lack the means for self-care.

Responsibility summersaults into a dead-end: members of society are expected to be accountable for shaping and maintaining a healthy natural and social environment, thus presumably improving the social determinants which nevertheless are of such structural rigidity that they conform to the brutal luck conditions that lie beyond personal influence and individual responsibility.

The role of citizenry was stressed in the Health Cities programs initiated in the 1980s. Improvement of urban living conditions is of utmost importance, but these programs would show better and more predictable results if they relied less on expert top-down advice that tepidly engages communities into participation and democratic decision-making. Regulations resulting from expert problem-solving often alienate segments of communities—the disabled, the aged, ethnic minorities— whose interests and needs remain ignored. Aseptic policies, often culture-insensitive, seek justice by placing themselves behind a context-free veil of ignorance, but real life is too complex to be satisfied by blanket dispositions that ignore "those who begin to develop a language and practices that acknowledge the complex interactions between power, knowledge, community, subjectivity and embodiment" (Petersen and Lupton 2000). Admittedly, this is a tall order, but should be seriously considered as public health loses its universality and is in danger of being replaced by a brand of liberal individualism that increases discrimination and inequality.

Redesigning the Public Health Agenda

The New Public Health has had peculiar effects on public health deliberation. Much of governmental responsibility toward preventive health care is being ignored as citizens are summoned to share and take over preventive measures. Instead of issuing regulatory preventive measures targeted at social and environmental contexts, public health concentrates its efforts on promotional and educational campaigns which, to the mind of traditional scholars, are unduly regulating the market. Much to the chagrin of conservative public health specialists, state intervention continues to be excessively normative, though rarely executive, on issues that they believe should be left to private initiative. As it has been evolving, they claim, public health has neglected the regulation of health risks at the same time increasing normative oversight of market transactions (Epstein 2003). Those advocating for a more robust State, insist that public health commitments are of the essence provided interventions are based on solid knowledge concerning causal disease mechanisms and efficient preventive measures.

Policy changes and economic constraints are not the only factors nurturing these transitions to the New Public Health. Emergent as well as reemergent diseases challenge traditional epidemiological strategies. HIV/AIDS could not be controlled by seclusion: measures such as isolation, quarantine, mass screening, and mandatory declaration serve no epidemiological purpose beyond statistical registration, at the same time unleashing social discrimination, moral rejection, job losses, and family disruptions. As social tension mounted, public health authorities admitted the need for an "exceptionalist approach" toward individuals rather than groups (Burr 1997). Tuberculosis, a scourge successfully controlled by mass screening, universal immunization, and effective medication, has reemerged as a fearsome public health problem that must be handled with due respect for individual patients' rights. Forced hospitalization was discussed and even carried out, but in-patient treatment could not be enforced under the provision that patients have a right to refuse treatment. Such diseases as Ebola, Hanta, West Nile virus are feared for their severity and high mortality of affected individuals, rather than for their impact on the population. Quite obviously, preventing and curing these conditions is seen as a public health problem, but its management has been promptly fitted into the professional competence of clinical medicine now managing prevention in all its stages.

The trend toward public health as community service planning and the new discipline called community medicine, seemed at first to bring public health policies nearer to the public, but this happened to be a mixed blessing, for community medicine "as a discipline increasingly experienced difficulties in defining its constituency and faced mounting problems of implementation in practice" (Warren 1997 in Porter 1999). Fragmenting political responsibilities creates enormous disparities between rich states, communities, or municipalities, and their poorer counterparts. Furthermore, decentralization erodes nationwide public health programs, so that epidemics which had been successfully curbed by federal governments, regained virulence after public health measures were put under local management, as experienced in Brazil with the resurgence of Dengue and Chagas endemics (Dias 2007).

The neoliberal tendency to dump health problems in the lap of individuals without considering their financial capacity to cope with self-sustained security and medical insurance reinstates a polemic that has plagued public health since early times. Nineteen century's medical ideology revived the concept that deficient personal health care—the proletariat's "poor health behavior" (Johanisson 1994 quoted in Porter 1999)—explains high levels of morbidity and premature mortality. Welfare States developed a strong governmental commitment to provide social security and medical care, but the conjunction of globalization, weakened States, and financial crises has contrived to reduce universally provided protection, allowing economic considerations to prevail over political and ethical commitments while neglecting sanitary needs.

Preliminary evidence is beginning to show that public health actions result in declination of preventable deaths, and that these mortality reductions "attributable to increases in public health spending are sizable, and may exceed the reductions achievable through similar expansions in local medical care resources" (Mays and Smith 2011). If confirmed, these findings would corroborate that traditional public health ought to recover its preventive agenda from its present capture by clinical practice.

The pervading "business as usual" perspective of liberalism is, of course, inhospitable to major reforms and to ideological confrontations where values and preferences are of more importance than the consistency of arguments. A historical illustration of this elusive point is given by Lord Beveridge's Report (1942), committed to law in 1946, advocating a national health service in Britain "which would cover each individual from the cradle to the grave," as compared to the tepid Health Care Reform in present day USA, with its development of prepaid medical care kindled by "easing costs of hospitalization rather than the idealistic notions about socialized medicine," as the concept of welfare is replaced by workfare (Porter 1999). Both the public and the private sector are plagued by economic constraints and difficulties in meeting the rising costs of medical care that cannot be squared by rationing or costs containment plans.

The issue is too polemic to allow a clear view, so perhaps the following, admittedly arbitrary, distinction may be employed: theory tends to amplify the agenda and concerns of a new approach to public health based on traditional values, whereas in practice, there is a general trend to reduce public health policies and programs and divert them from a governmental population-wide concern to a personal matter of self-responsibility anchored in medicine's clinical practice, and running under the name "New Public Health." In other words, the gap between proclamation and actual practices is deepening, as the public sphere tends to unload tasks and responsibilities on the private realm, thus increasing the gradient between the empowered and the destitute.

Public health is rife for the development of its own brand of ethics. Although bioethics appears as the most plausible candidate for this undertaking, some scholars have embraced the opposite view, believing bioethics to be so utterly committed to singular medical situations and to the clinical encounter, that it appears irredeemably inappropriate to approach the issues of collective health and disease problems (Bayer and Fairchild 2004). Subsequent chapters hope to prove otherwise, by probing the ethical issues raised by the major changes affecting public health.

References

Armstrong, D. (1995). The rise of surveillance medicine. *Sociology of Health & Illness, 17*, 393–404.

Arouca, S. (2003). *O dilemma preventivista*. São Paulo, Rio de Janeiro: Editora UNESP & Editoria FIOCRUZ.

Authority, P. H. S. (2011). *Towards reducing health inequities*. Vancouver: Population & Public Health, Provincial Health Services Authority.

Bayer, R., & Fairchild, A. L. (2004). The genesis of public health ethics. *Bioethics, 18*(6), 473–492.

Burr, C. (1997). The AIDS exception: Privacy vs. public health. The Atlantic Monthly(June): 57–67.

Dias, J. C. P. (2007). Globalization, inequity and Chagas disease. *Cadernos de Saúde Pública, 3*(Suppl. 1), S13–S22.

Epstein, R. A. (2003). Let the shoemaker stick to his last: A defense of the "old" public health. *Perspectives in Biology and Medicine, 46*(3 Suppl), S138–S159.

Fee, E. (1993). Public health, past and present: A shared social vision. In G. Rosen (Ed.), *A history of public health*. Baltimore: The Johns Hopkins University Press.

Gostin, L. O., & Bloche, M. G. (2003). The politics of public health: A response to Epstein. *Perspectives in Biology and Medicine, 46*(3 Suppl.), S160–S175.

Judt, T. (2010). *Ill fares the land*. New York: The Penguin Press.

Lalonde, M. (1981). *A new perspective on the health of Canadians*. Ottawa: Minister of Supply and Services.

Lazzarato, M. (2009). Neoliberalism in action. Inequality, insecurity and the reconstitution of the social. *Theory, Culture & Society, 26*, 109–133.

Lupton, D. (2003). *Medicine as culture*. London: Sage Publications Ltd.

Mays, G. P., & Smith, S. A. (2011). Evidence links increases in public health spending to declines in preventable deaths. *Health Affairs (Millwood), 30*(8), 1585–1593.

McKeown, T. (1979). *The role of medicine*. Oxford: Basil Blackwell.

Petersen, A., & Lupton, D. (2000). *The new public health*. London: Sage Publication Ltd.

Porter, D. (1999). *Health, civilization and the State*. London: Routledge.

Rose, G. (1985). Sick individuals and sick populations. *International Journal of Epidemiology, 14*(1), 32–38.

Vandenbroucke, J. P. (1994). New public health and old rhetoric. *BMJ, 308*(6935), 994–995.

Wiley, L. F. (2010). Mitigation/adaptation and health: Health policymaking in the global response to climate change and implications for other upstream determinants. *The Journal of Law, Medicine & Ethics, 38*(3), 629–639.

Chapter 4
Bioethics in Public Health

Abstract The need to adapt clinical bioethics to the issues raised by public health policies and interventions did not become obvious till the later 1990s. The interpersonal principlist approach was inappropriate for the social concerns of public health, although autonomy, beneficence, nonmaleficence, and justice had to be incorporated and adapted to collective scenarios. Autonomy touches on the basic concern of developing public policies that avoid undue limitation of individual free will, and justice is of major import when reflecting on inequality and the allocation of scarce resources. Interventions designed to prevent disease and promote health are expected to be beneficial at the population level, and care must be taken to respect human rights in the understanding that basic rights and public health are mutually supportive.

Public health bioethics has up till now developed along the general lines of responsibility, prevention, and precaution. Responsibility is a core value, but not easy to determine when agents are impersonal institutions and accountability cannot be rendered to anonymous collectives. Prevention is basically a technical endeavor, required to be effective—problem-solving—and efficient—sustainable cost/benefit ratio—its ethical component addressing the just distribution of preventive policies. Precaution refers to decision-making in uncertainty, its ethical value having often been overshadowed by strategic negotiations of vested interests.

Keywords Autonomy • Human rights • Precaution • Prevention • Principlism • Responsibility • Social bioethics

Public health in its contemporary form is struggling to define and articulate its core values. In search of a "framework which expresses fundamental values in societal terms," J. Mann proposed to bind the parallel efforts of promoting and protecting both human rights and health in order to secure fairness and avoid discrimination, thus identifying the role of public health in opposing human rights violations (Mann et al. 1994; Mann 1997). On the basis of this general approach, scholars began to take notice that public health was in urgent need of an ethical approach.

M. Kottow, *From Justice to Protection: A Proposal for Public Health Bioethics*,
SpringerBriefs in Public Health 1, DOI 10.1007/978-1-4614-2026-2_4,
© Miguel Kottow 2012

Public health and bioethics had their first encounter in the 1990s, when the proceedings of an "Industrial Epidemiology Forum's Conference" devoted to professional ethics for epidemiologists were published (Fayerweather et al. 1991). This initiative, however, failed to create a sustained interest in the ethics of public health. In the heyday of clinical bioethics, public health was seen as "bedeviled by moral issues" (Daniels and Sabin 1998) and "the field" was considered to be "riddled with serious conceptual and ethical problems" (Williams 1984, quoted in Skrabanek 1990).

A number of important papers proposing general frameworks for public health ethics have been published, producing basic outlines of great interest for a nascent discipline by stressing values such as effectiveness, relevance to community needs, cost/benefit analysis, political feasibility, and public accountability (Callahan and Jennings 2002; Childress et al. 2002; Kass 2004). A plea for a more comprehensive bioethics agenda showed concern for the quandaries of resources allocation and the influence of social determinants of health, indulging in theoretical probing too far removed from practical concerns to have a lasting influence (Brock 2006). The problem arising from the opposition between urgent attendance of essential need, versus political and economic considerations, remains open as indeed it inevitably will when rigid external circumstances are allowed to shape and determine public health programs. Some unsolvable contradictions appear when one framework centers around social justice while another signals its wariness because justice, like autonomy, might go astray and unduly override all other factors (Baum et al. 2007).

Public health bioethics must deal with a brand of principlism that evolved, for better or worse, as an interpersonal form of ethics mainly concerned with the clinical encounter and the physician–patient relationship, but ill-equipped to face the challenges presented by the collective issues of public health which differ substantially from those of clinical medicine. The Georgetown principlist school having successfully developed a bioethics doctrine based on four principles—autonomy, beneficence, nonmaleficence, and justice—was soon contested by critics who considered them as moral check-lists rather than action-guiding principles. Although the four principles of doctrine are more relevant to clinical ethics, they cannot be dismissed in approaching the bioethical quandaries of public health precisely because, acting as check-lists, these moral tenets must always be incorporated into ethical reflection, nuanced to find their place in different cultural environments and adapted to the practices they address.

Autonomy

Mill's original proviso, that liberty should only find its limits when interfering with the liberty of others, is insufficient to deal with social issues in the context of public action. Autonomy is an anthropological attribute to be distinguished from its actual exercise which is more often than not determined by external factors. Inasmuch as normally developed individuals are intrinsically autonomous, any limitations in the exercise of autonomy are externally caused, constituting an imposition that acts as a brute fact for which those affected cannot be held responsible.

Autonomy exists by default in normal, mature human beings, and observers must be extremely cautious in diagnosing defective autonomy, its degree, and limits. The autonomy of children cannot be scaled by age, nor can a general deficiency be imputed to individuals with senile deterioration, who may lucidly decide in some circumstances even though they are pitifully inadequate in others. The idea of relational autonomy suggests that freedom of choice is a social construct that functions in the political and social environment we all share (Baylis 2008). Therefore, the practice of public health will engage individual autonomy in a cooperative drive to serve the common weal. Or, in more realistic terms, individuals should not claim autonomy to interpose obstacles in the path of public health programs that are genuinely involved in benefitting the social good.

Beneficence and Avoidance of Harm

Beneficence and nonmaleficence gain a particular flavor in the context of social ethics. Beneficial public health actions are usually presented as having positive effects on populations at large, while harm befalls individuals: a vaccination program will improve the population's resistance to an epidemic, but it will be isolated individuals who may suffer the ill-effects of unwanted side reactions. When evaluating benefit/risks ratios of, say, an epidemiological investigation, researcher must bear in mind that potential benefits, if any, will diffuse through the collective or obtain for privileged groups, whereas risks will burden individual participants. Paraphrasing Rose's preventive paradox, what might benefit the population may have little or even detrimental effect on individuals.

Justice

Justice has also come to be understood as a relational ethical dimension, suggesting that it functions as two concepts in public health: as distributive justice and as social justice. Distributive justice applies to the allocation of finite, tangible health-related goods and services, whereas social justice is a general appeal focused on fair access to rights, opportunities, power, and self-respect. Both aspects interlock, as shown by the concept of social injustice that combines domination—disrespect for rights and opportunities—with deprivation—unfulfilled needs of essential goods (O'Neill 2002; Young 2011). In any event, justice in public health issues is always comparative and relational, it is invariably embedded in social reality and therefore must specify its reach and commitment.

"Social justice is concerned with human well-being," and one way of understanding well-being is by compiling a list of its elements: health, personal security, opportunities to exercise practical reasoning, respect, attachment, and self-determination.

In this view, public health is about improving human well-being by strengthening health and focusing on "special moral urgency to remediating the conditions of those whose life prospects are poor" (Powers and Faden 2006). This very plausible train of thoughts is somewhat put to question when these authors go on to assert that social justice is not only about combating deprivation, but also includes exercising comparative well-being for everyone. As repeatedly noted, it is hard to see public health as a social force capable of embracing holistic goals of global socioeconomic equality. The claim of a universal and egalitarian rights to health care is feeble and ineffectual when compared with identifying obligations to act at the national level which, in the case of medical care, requires specification of "*who* has to do *what* for *whom*" (O'Neill 2002, emphasis in the original).

A strong reason for replacing justice criteria when approaching inequalities in health is that universal, egalitarian access to health care may have negligible, even negative effect on the social gradients and disparities of health indicators, allowing the affluent and better educated to live longer and healthier lives. Egalitarian health care does not guarantee that individuals will get the medical services they require, it only means that they will get the same care everybody else is receiving. One of the main topics of bioethics in public health is allocation and rationing of scarce resources, which cannot be solved in terms of fairness because rationing inevitably means neglecting the legitimate claims of those who will be turned down if they do not meet whatever selection criteria are being upheld. Furthermore, economists have problems allocating scarce fiscal resources because they initially fall back on individual evaluation criteria, such as QALY, DALY, TTO, and adjust them to collective programs.

Beyond Principles

Theorists analyzing ethical principles vary from "defending a particularist conception" which "sees little if any role for moral principles" (Dancy 2004), to those considering that commitment "to a plurality of normative principles…does not undermine, but rather provides a basis for practical judgments" (O'Neill 2009). Practical reasoning may well conclude that the issue of principles may be of theoretical interest, yet lacks relevance in resolving moral dilemmas or clarifying ethical problems.

Authors of ethical frameworks try to remain flexible and open-minded, suggesting that their "foundational principles and values" are applicable in "numerous ways," and admitting that they may entail "conflicts among general moral considerations." Ethics may be too widespread and theoretical to be applicable in practice. Inasmuch as some frameworks overshoot into global dimensions or bask in proposing unlimited goals of justice and well-being, they lose credibility when widening the gap between theory and praxis. To the contrary, ethical norms may be conceived too narrowly to accommodate diverse perspectives which, finally, become weakened by contextual specifications. Like principles, frameworks raise important issues in public health ethics but offer little in the way of guidance.

In spite of the short-comings of principle-based ethics, and the unending polemics on the theoretical solidity and practical value of principles, repeated efforts have been spent in presenting a principle-based ethics for public health. A tetrad of principles tailored for public health has been suggested and recommended for its heuristic and justificatory value: the harm principle, the principle of least restrictive means, the reciprocity principle, and the transparency principle (Upshur 2002). Fatigue with discussions over ethical principles may explain why this initial suggestion remained unheeded and uncontested, probably because its maxims are too general to be of specific use to public health.

Expanding the Bioethics Agenda

As bioethics developed into a substantial academic discipline, it has had to face accusations of neglecting its commitment to practical reason, thereby developing its own brand of an ethical 90/10 gap: most of its attention is devoted to issues that only concern the more affluent minority of the world's population (Rennie and Mupenda 2008). "Sociologists and social epidemiologists challenge bioethicists, especially those working in developed countries, to be socially and culturally relevant" (Azetsop and Rennie 2010). This sound advice is to be complemented by a reciprocal relationship where bioethics infuses ethical as well as cultural values and attitudes into the practices of epidemiologists and public health policy-makers. Research and practice in public health ought to commit themselves to upstream their efforts, taking the most urgent and basic health and medical needs as their point of departure (Whitehead et al. 1998).

Bioethics as it had developed for 30 years lacked the span to tackle problems presented by issues concerning the environment, nonhuman life and, in general, the values involved in collective actions. Although the final aim of health protection is to assist and safeguard the individual human being, public health is instrumentally and strategically committed to perform at the level of populations, seeking epidemiological knowledge and proposing health care measures that can only be achieved at the social level. Individuals cannot repel epidemics, nor can they mitigate environmental deterioration, and yet they do suffer the ill-effects of misfired public actions. Bioethics in public health must deal with collective interventions, their effects on individuals, and the repercussion of individual actions and attitudes on social health programs. Politics and ethics share the tension of addressing the collective but finally acting on individual human beings.

In a more general analysis, four major dimensions inspiring public health orientation can be identified: responsibility, prevention, precaution, and protection.

Responsibility

No ethical perspective can dispense with responsibility, and all human actions must ultimately respond to ethical scrutiny. Responsibility is inherent to any human act that affects others, therefore being subject to ethical evaluation, so that designing an ethics of responsibility seems a redundancy. Ethical responsibility may be preceded by technical, economic, professional, or some other form of accountability that evaluates performance of tasks, fulfillment of roles, or pursuance of goals but, finally, ethical assessment cannot be dispensed with. Pragmatic responsibility usually answers to an existing normative code, actions being judged by their fidelity to, or neglect of, established guidelines that regulate specific activities. In contrast, ethical responsibility is not code-bound, for ethics is deliberative rather than rigidly prescriptive; its norms are recommendations but not impositions, lest an ethical doctrine become moral dogma.

A mutual responsibility develops between public health and the citizenry, flourishing in the twentieth century under the banner of socialism in the now defunct Second World and, in a more liberal form, in the Social States developed in post-World War II Central Europe: government is responsible for disease prevention, health promotion and, to a great extent, medical care of the population especially in countries with a comprehensive national health service. For their part, individuals are called upon to comply by disciplinary measures required for effective preventive campaigns, and to have a percentage of their income retained by the State to finance national health expenditures and medical coverage for the population. This very broad description varies from one nation to another, and depends on externalities—economics, technological development, cultural shifts—but, all in all, it illustrates how traditional public health developed between the post-westphalian creation and strengthening of nation-states, and the advent of the "new public health" in the second half of the twentieth century under the predominant influence of global economic interests.

Responsibility goes hand in hand with accountability. In health matters, accountability occurs at the political level, where expenditures and goal assessment are publicly presented, at the institutional level measuring effectiveness, and at the end-performance where health-care providers and recipients evaluate satisfaction with results obtained (Daniels and Sabin 1998). Since the end-point of health policies and actions is elusive and subject to any number of criteria, accountability in this area lacks objectivity and is prone to embellished institutional reports or to biased dissatisfaction by those receiving benefits short of their expectations (Klein 1982). In a truly participative society, legitimate accountability ought to be rendered at the grass-root level.

In the realm of public health, opinions vary between those who require individuals to account for their personal performance, eventually speaking up to defend their interests, and views that require organizations to carry an institutional, collective burden of responsibility. A compromise suggestion postulates that public health organizations are moral agents in a way that differs from the moral agency of individuals (Coughlin 2009). In sum, political, institutional and, generally speaking, collective responsibility in public health is a necessary element of the broad picture of ethics, but will not be specific and sufficient to evaluate political orientations and particular practices aimed at disease prevention and health care at the population level.

A less explored link between public health and responsibility is established by resorting to virtue ethics. Virtue theory is invoked to evaluate that the right motivations inspire actions of responsible scientific attitudes in epidemiologists (Weed and McKeown 1998). Virtues are essential to restore public trust in political action including public health care (Coughlin 2009), and the American College of Epidemiology Ethics Guidelines for Epidemiologists links virtuous conduct with professional obligations, that is, with carrying out their social roles in a responsible manner. A virtuous character will lead epidemiologists to be prudent and responsible in broaching polemic issues, especially when serving in advisory capacity to public health policies (Holland 2007).

Public health is an activity anchored in institutions dealing with policies and collective actions aimed at health issues at the population level, making it difficult to apply virtue ethics which are essentially attributable to individual human beings. Virtuosity of public health physicians as they exist in the British health system can be envisioned in a more straightforward way then in the case of officials anonymously enmeshed in institutional organizations. British public health physicians are under permanent stress in their mediator function between individual rights and the responsibilities of society for all its members, having to virtuously balance the excellence of their services against "managerial and medical orthodoxy" (Horner 2000). Alas, virtue ethics is a fragile and understaffed vessel.

Prevention

Public health and prophylactic activities have been so firmly associated that "preventive medicine" is often used as a synonym for public health. Whatever else it might be, public health essentially takes primary prevention of disease as its major goal.

Traditional prevention recognized specific threats to health and throve to present solidly proven and technically feasible solutions. Prevention in public health is basically a pragmatic approach with the moral obligation of applying its programs equitably, thus fulfilling a social obligation to prevent disease in an efficient, timely, and fair fashion. Ever since clinically based individual Preventive Medicine took over, hyperactivity has led to premature and scientifically immature interventions, creating a host of moral problems that ensue when prevention becomes a market commodity plagued by uncertain risks and dubious benefits (Williams 1984; Starfield et al. 2008), raising the question "Why is preventive medicine exempted from ethical constraints?" (Skrabanek 1990).

The fusion of primary and secondary prevention has radically changed from its original aim of securing productivity by protecting the working population, to contemporary medicalization of "healthy" individuals in whom some statistically deviant lab values are detected. Clinically based prevention avidly explores the human body, elevating any anomalous or atypical finding into the rank of disease risk to be actively treated with medication that is often far from effective and by no means always harmless. The medical establishment and pharmaceutical companies make huge profits by promoting frequent check-ups, extensive diagnostic probing, wide open registries of predispositions, preclinical conditions, and risk factors. Data are manipulated into clinical significance, leading to overmedication, exhortations to follow healthy live-styles, the promotion of dietary products, sports, and home-training gadgets. As an anonymous wit put it, "to follow all the health care suggestions would require a 29-h long day." An exaggeration that illustrates how self-care and self-responsibility can easily lead to self-blame for neglecting to comply with full-fledged health care programs (Verweij 1999).

Prevention Might be Overrated

To a large extent, individual prevention in health becomes a commodity when public health relinquishes its traditional agenda of keeping disease away from populations. As prevention moves away from inspiring public action and nestles in the realm of individual health care, it ceases to be a safeguard for collective health. Self-care is monitored by physicians as part of the clinical encounter, hastily being absorbed by indications for early onset therapeutics. This trend, though not universal, is gaining momentum and influence, tending to reduce public health funds, and slowing the pace of societies and institutions intent on providing universal medical services. European countries that for decades had been at the forefront of comprehensive national health policies and medical care plans, are now denying these services to all but regular citizens (Fassin 2001), reducing insurance coverage, and increasing co-payments for medical acts including prescriptions. The biomedical scientific community has also neglected the investigation of vaccines; according to the WHO's International Clinical Trials Registry WHO's, disease prevention studies gets an infinitesimal part of funds allocated to research, the lion share going to clinical trials (Karlberg 2010).

A conflict "between economics and medical morality is raised in the current interest in reallocating funds from curative to preventive medicine… When they [the healthy] become ill and are in immediate need, however, they are far less likely to think about the economic impact of care versus prevention." This is quite in line with "the traditional canons and expectations of medical morality demand[ing] help for those who are ill now and present themselves for care" (Pellegrino and Thomasma 1981). This primeval text on the philosophy of medicine requires that moral demands of immediate medical care trump over economic considerations and provide actually needed medical attention even if resources must be siphoned from the public prevention programs. The question whether early disease prevention is preferable to the burdens of curative treatment does not enjoy such a straightforward affirmative answer as often claimed. Insufficient information, poor advise, or increased risk perception, may unleash costly preventive strategies that could have been avoided.

Efforts at restituting health prevention to its proper place as a public health activity face accusations of State paternalism. Wary of appearing to support authoritarian stances, juridical procedures have often solved conflicts in favor of individual autonomy at the price of being detrimental to collective interests. The pendulum may be swinging too far away from social commitment, for public health should resist individual exceptionalism by supporting its activities with scientific and ethical soundness. Public health belongs in the "realm of the political and the ethical" and "is also a practical science" seeking "reasonable and practical means of... limiting liberty to promote the health of the public generally" (Beauchamp 1985).

Public health has lost great part of its traditional tuition over disease prevention. Prevention is steeped in uncertainty ever since the opaqueness of multicausality has unfocused well-targeted cause–effect intervention, clearing the way for promotion and precaution, both strategies pretending to be adequate substitutes for decision-making in unknowingness. Since most risk factors are based on associations rather than on determinate causal factors, prevention and promotion lack strong convincing powers, and the so-called precautionary principle has shown itself too maneuverable by vested interests to be a reliable ethical defender of the public weal (see Chap. 8).

Precaution

Precaution has often been presented as a special form of prevention to be applied when scientific evidence is insufficient to orientate action. Both strategies are expected to avert harm or reduce the risks of unwanted events, but they differ in their approach. Prevention, as previously discussed, is based on a clear understanding of an unwanted situation and the availability of effective means to neutralize or mitigate its deleterious consequences. Precaution, on the other hand, is resorted to when the magnitude of a potential or actual harmful situation cannot be accurately estimated, and scientific evidence is insufficient to decide if and how the detected or anticipated problem should be handled. The mark of prevention is effectiveness based on deterministic causal evidence, whereas that of precaution is decision in uncertainty using estimates and assessment to evaluate probabilities. To be sure, "precaution requires more and better science" (Martuzzi 2007), but as uncertainty is reduced, prevention substitutes for precaution. Strategies such as discounting evidence, reducing causal criteria, weakening rules of evidence, and adapting levels of significance have been advanced in order to improve the ductility of the precautionary approach (Weed 2004).

A so-called precautionary principle was presented in Europe in the 1980s, in an effort to provide ethical orientations in public health crisis that could not be rationally handled. The most obvious cases of governmental disorientation were the sudden appearance of prion-induced Creuzfeld Jacob disease presumed to be transmitted from sick cows, and the large-scale production of genetically modified organism including food crops such as maize, soya, wheat, and tomatoes. In all these cases, strategies posing as precautionary had the unintended effect of increasing uncertainties, and revealing their vulnerability to erroneous interpretation and interested manipulation.

Possibly, the most illustrative failure of a precautionary policy was enacted when the World Health Organization panicked and declared a H_1N_1 flu pandemic, based on deceptive expert information (2008–2010). Countries that complied by these precautionary exhortations spent enormous amounts of money stocking vaccines that were never needed because the pandemic had been maliciously overrated by experts sharing financial interests with the vaccine producing industry.

Current definitions of precaution are unable to describe a complex and polemic concept that has been elevated to the rank of principle. The Rio Conference of 1992 declared: "When there are threats of serious or irreversible damage, lack of full scientific certainty shall not be used as reason for postponing cost-effective measures to prevent environmental degradation." An editorial in *Lancet* specifically indicates: "Where there are significant risks of damage to the public health, we should be prepared to take action to diminish those risks, even when the scientific knowledge is not conclusive, if the balance of likely costs and benefits justifies it" (Horton 1998). The point is that if appropriate cost/benefit metrics could be applied, the uncertainties of precautionary attitudes would be advantageously replaced by well-substantiated prevention.

In view of the widespread invocation of precaution in such scenarios as the environment, public health, cutting-edge biomedical research, and other fields of social import, more clarity is required and may be gained by making a distinction between precaution *ex ante,* and its *ex post* mode. Anticipatory precaution refers to innovations, mostly technical in nature, that require careful assessment as to the benefits being offered, the known and suspected risks, as well as the distribution of positive and negative effects on the social and natural environment. Major points of contention are the eventual long-term effects that may jeopardize future generations. *Ex ante* precaution has been accused of crippling scientific progress and technical improvements, but its backers point at the negative aspects of uncontrolled innovative hubris. To the contrary, it is argued, innovations herald improvements that will change circumstances for the better. Introducing new pharmaceutical agents, as well as implementing immunization strategies when epidemics or even pandemics threaten, is a salient example where precaution has been extensively and hotly debated, at times being applied with disastrous after effects.

Environmental polemics centers on ongoing processes that are causing climate changes, toxicities, and degradations, affecting human quality of life and threatening biodiversity, where remedial *ex post* precaution is urgently suggested. Public health is also faced with situations where tolerating a *status quo* produces negative effects that ought to be curtailed, as in the widespread and far from innocuous use of agrochemicals.

In the pharmaceutical market, populations have been exposed to drugs that were insufficiently studied and had to be recalled after having caused considerable damage, the talidomide catastrophe being a pungent case of thousands of malformed children born of mothers who had received a tranquilizer, the teratological effects of which had been insufficiently studied. As the pharmaceutical industry feels urged to innovate for the sake of securing patents and profits, drugs are prematurely put on the market and have to be recalled after severe and often lethal effects occur. In recent years, authorized and widely distributed anti-inflammatory and metabolism-stabilizing drugs had to be hastily recalled after causing severe complications.

Public health policies, like environmental plans, have been frequently overrun by potent interests that impose their strategies under the mantle of an ethics of precaution.

Nuclear disasters, oil spillages, pharmaceutical drugs, and car models that had to be recalled are just a few examples of unexpected risks that plague technocracy. Mobilizing or assuaging fear often relies on scanty information, and vested interests have been known to distort information to make their point and win adepts.

The abundant literature dealing with the precautionary principle is beyond the reach of public health ethics, but there is sufficient evidence that precaution is not so much an ethical issue as a negotiation between powerful parties defending their interests. The power of big business usually outwits the fears and caution of democratic forces, prudence and caution being overpassed by pragmatic, market-oriented interests. In fact, it can be said that precautionary tactics rarely comply with the ethical requirement of submitting to democratic deliberation (Callon et al. 2001). As things stand, the precautionary principle is a strategic instrument operating at the service of potent influences that suffer scanty opposition in pursuing their goals because the democratic participatory element that would control the ethical solvency of precaution is all but absent or easily circumvented (Goldstein 2007).

References

Azetsop, J., & Rennie, S. (2010). Principlism, medical individualism, and health promotion in resource-poor countries: Can autonomy-based bioethics promote social justice and population health? *Philosophy, Ethics, and Humanities in Medicine, 5*, 1.

Baum, N. M., Gollust, S. E., Goold, S. D., & Jacobson, P. D. (2007). Looking ahead: Addressing ethical challenges in public health practice. *The Journal of Law, Medicine & Ethics, 35*(4), 657–667. 513.

Baylis, F. (2008). Choosing a path: Setting a course for the journey. *The American Journal of Bioethics, 8*(12), W4–6.

Beauchamp, D. E. (1985). Community: The neglected tradition of public health. *The Hastings Center Report, 15*(6), 28–36.

Brock, D. W. (2006). Etjical issues in the use of cost effectiveness analysis for the priorisation of health care resources. In S. Anand, F. Peter, & A. Sen (Eds.), *Public health, ethics, and equity* (pp. 201–223). Oxford: Oxford University Press.

Callahan, D., & Jennings, B. (2002). Ethics and public health: Forging a strong relationship. *American Journal of Public Health, 92*(2), 169–176.

Callon, M., Lascoumes, P., et al. (2001). *Agir dans un monde incertain.* Paris: Éditions du Seuil.

CIOMS. (2002). *International ethical guidelines for biomedical research involving human subjects. C. f. I. O. o. M. S. (CIOMS).* Geneva: CIOMS/WHO.

Coughlin, S. G. (2009). *Ethics in epidemiology & public health practice.* Washington: American Public Health Association.

Childress, J. F., Faden, R. R., et al. (2002). Public health ethics: Mapping the terrain. *The Journal of Law, Medicine & Ethics, 30*(2), 170–178.

Dancy, J. (2004). *Ethics without principles.* Oxford, New York: Oxford University Press.

Daniels, N., & Sabin, J. (1998). The ethics of accountability in managed care reform. *Health Affairs (Millwood), 17*(5), 50–64.

Fassin, D. (2001). Quand le corps fait loi. La raison humanitaire dans les procédures de régularisation des étrangers. *Sciences Sociales et Santé, 19*, 5–34.

Fayerweather, W. E., Higginson, J., et al. (1991). Ethics in epidemiology. Industrial Epidemiology
    Forum's Conference. *Journal of Clinical Epidemiology, 44*(1), 41S–50S.
Goldstein, B. D. (2007). Problems in applying the precautionary principle to public health.
    Occupational and Environmental Medicine, 64(9), 571–574.
Holland, S. (2007). *Public health ethics.* Cambridge: Polity Press.
Horner, J. S. (2000). For debate: The virtuous public health physician. *Journal of Public Health
    Medicine, 22*(1), 48–53.
Horton, R. (1998). The new public health of risk and radical engagement. *Lancet, 352*(9124),
    251–252.
Karlberg, J. P. E. (2010). WHO International Clinical Trial Registry platform - An in-depth analy-
    sis-. *Clinical Trial Magnifier, 3*, 333–344.
Kass, N. E. (2004). Public health ethics: From foundations and frameworks to justice and global
    public health. *The Journal of Law, Medicine & Ethics, 32*(2), 232–242. 190.
Klein, R. (1982). Performance, evaluation and the NHS: A case study in conceptual perplexity and
    organizational complexity. *Public Administration, 60*, 385–407.
Levine, C., Faden, R., et al. (2004). The limitations of "vulnerability" as a protection for human
    research participants. *The American Journal of Bioethics, 4*(3), 44–49.
Mann, J. M. (1997). Medicine and public health, ethics and human rights. *The Hastings Center
    Report, 27*(3), 6–13.
Mann, J. M., Gostin, L., et al. (1994). Health and human rights. *Health and Human Rights, 1*(1),
    6–23.
Martuzzi, M. (2007). The precautionary principle: In action for public health. *Occupational and
    Environmental Medicine, 64*(9), 569–570.
McCormick, J. (1994). Health promotion: The ethical dimension. *Lancet, 344*(8919), 390–391.
O'Neill, O. (2002). Public health or clinical ethics: Thinking beyond borders. *Ethics & International
    Affair, 16*(2), 35–45.
O'Neill, O. (2009). Applied ethics: Naturalism, normativity and public policy. *Journal of Applied
    Philosophy, 26*, 219–230.
Pellegrino, E. D., & Thomasma, D. C. (1981). *A philosophical basis of medical practice.* New
    York, Oxford: Oxford University Press.
Powers, M., & Faden, R. (2006). *Social justice.* Oxford, New York: Oxford University Press.
Rennie, S., & Mupenda, B. (2008). Living apart together: Reflections on bioethics, global inequal-
    ity and social justice. *Philosophy, Ethics, and Humanities in Medicine, 3*, 25.
Skrabanek, P. (1990). Why is preventive medicine exempted from ethical constraints? *Journal of
    Medical Ethics, 16*(4), 187–190.
Starfield, B., Hyde, J., et al. (2008). The concept of prevention: A good idea gone astray? *Journal
    of Epidemiology and Community Health, 62*(7), 580–583.
Upshur, R. E. (2002). Principles for the justification of public health intervention. *Canadian
    Journal of Public Health, 93*(2), 101–103.
Verweij, M. (1999). Medicalization as a moral problem for preventative medicine. *Bioethics, 13*(2),
    89–113.
Weed, D. L. (2004). Precaution, prevention, and public health ethics. *The Journal of Medicine and
    Philosophy, 29*(3), 313–332.
Weed, D. L., & McKeown, R. E. (1998). Epidemiology and virtue ethics. *International Journal of
    Epidemiology, 27*(3), 343–348. discussion 348–349.
Whitehead, M., Scott-Samuel, A., et al. (1998). Setting targets to address inequalities in health.
    Lancet, 351(9111), 1279–1282.
Williams, G. (1984). Health promotion—Caring concern or slick salesmanship? *Journal of
    Medical Ethics, 10*(4), 191–195.
Young, I. M. (2011). *Responsibility for justice.* Oxford, New York: Oxford University Press.

Chapter 5
Ethics of Protection I: Theoretical Fundaments

Abstract Ever since Hobbes, and throughout all political philosophies, nation-States are expected to honor the obligation of securing personal, patrimonial, and territorial protection to their subjects. Even libertarian minimal State proponents do not exempt the ruling powers from protecting their citizenry. Philosophers have put protection at the ethical basis of interpersonal relationships, as illustrated by care due to the newborn, future generations (Jonas), and the defenseless Other (Lévinas).

Complex societies are bound to provide protection beyond physical harm and patrimonial dispossession. Historically, nation-States have extended their responsibility to provide protection against catastrophes and biological threats such as epidemics and pandemics, and yet have been reticent in providing public medical care.

Protection goes beyond the ethics of care that is committed to proximal aid within familial and neighborly relations, for protection is equally due to the marginalized, the disempowered, and the distant destitute. Poverty not being a natural condition, has historical roots of dominance, colonialism, exploitation which need to be repaired.

Protection aims to help develop personal and collective autonomy, and should not be confused with paternalism which supplants and often disregards autonomy, especially in its unjustified authoritarian form. Protectionism, which is a policy aimed at safeguarding interests, is also distinct from an ethics that aims at protecting the weak.

Keywords Autonomy • Ethics of care • Paternalism • Protectionism • Protective States

The output of philosophical ethics is sparse, and the "language of morality" is in "grave disorder" (MacIntyre 1984). Ethical theory for applied ethics, including bioethics, has not provided a coherent background on which to erect plausible guidelines for social practices capable of producing the external goods required by society without neglecting the internal goods of excellence in performance and ethical probity

M. Kottow, *From Justice to Protection: A Proposal for Public Health Bioethics*,
SpringerBriefs in Public Health 1, DOI 10.1007/978-1-4614-2026-2_5,
© Miguel Kottow 2012

(Beauchamp 2004). Unsubstantiated proclamations insist that we currently live in a better world than ever before, where justice and respect of human rights are, perhaps slowly but surely, evolving toward fulfillment. Empirical evidence leads to a more pessimistic reading of reality, reporting that inequities and disempowerment are on the rise. The complacence of "high-minded charters" and "powerfully worded statutes, conventions, and treaties" has a paralyzing effect on initiatives seeking change (Farmer 1999). The often presented argument that major issues may be viewed as a half-empty or half-full glass is not appropriate, for the full half refers to important but nonessential benefits for the well-off, whereas the empty half addresses unmet basic needs, neglect and widespread suffering that require urgent attention.

Traditional approaches proclaiming universally valid ethical principles that emphasize human rights—United Nations 1948—social, economic, and cultural equality—United Nations 1966—and social bioethics—UNESCO 2005—are systematically ignored or honored in the breach by rogue states as well as by respected signers of these institutional documents. Their lack of impact has prompted institutions such as communities, universities, NGOs, and others to effectively broach and solve specific problems at the local level. Clearly designed projects have been successful within the limits of particular programs, somewhat less so in long-lasting effects, since problem-solving prevails over structural changes and the setting of precedents.

The bioethical groundwork that will allow gaining new perspectives and a more reliable coverage of health care and medical assistance to those in dire need has yet to be undertaken. Previous suggestions, like P. Farmer's commitment to provide medical assistance to those whose human rights have been ignored or violated, is certainly a crucial and admirable step in the right direction (Farmer and Campos 2004). It needs to be stressed, though, that ethical concern and practical assistance must also reach those who are, as matters stand, too radically excluded to claim human rights, the voiceless and disempowered who are beyond the reach of helping hands. The ethics of protection is a tentative effort to endorse deliberation in search of such a general ethical mantle, hopefully capable of withstanding critical probing and be supportive to those engaged in practical action. Assisting and protecting the underprivileged in health and medical matters ought to be a systematically anchored primary task of public health.

Roots of Protection

"Nature hath made men so equal," wrote Hobbes in 1651, that tending "to desire the same things" they would employ violence "to obtain gain," thus living in permanent condition of war where "right and wrong, justice and injustice have no … place." Such a life, as often quoted, would be "solitary, poor, nasty, brutish, and short," and can only be avoided by resorting to the "covenant of every man to every man," creating and submitting to a central "sovereign" power, "on confidence to be *protected* by him against all others" (Hobbes 1978). Personal and patrimonial protection is the first, and for libertarians the only obligatory function of the State. An ultraminimal State that might frown on levying the least taxes required to guarantee basic universal personal protection is untenable, because it is "impermissible" for any State to fail in "providing protective services for all" (Nozick 1980). Under any conceivable political sovereignty, individual and patrimonial protection remains essential. Protecting life, limb and purse of its citizens is the one and only obligatory function of the libertarian State, hence the universal acceptance and support of a military force for territorial protection, and a police force for personal security and social order.

Anchoring personal protection as a fundamental obligation of the State is a necessary but insufficient political framework, tending to fall into a fallacious train of thoughts that confides in an invisible hand that will provide the recognition, respect, fairness, and security that each and every human being requires as an equal member of the species but also as singular individual. Neither biology nor philosophy, and certainly neither ethics nor religion, deny that personal traits make human beings unique. Human plurality consists of sameness and singularity (Arendt 1958). When setting the stage for social medicine, Virchow wrote in 1849: "Health and disease are, of course, always the possession of individuals, for life is realized en each singular being, not in the mass" (Deppe and Regus 1975). This uniqueness has been employed to question egalitarianism because it denies a person the "solidity of his own sense that he is real…and alienates people from themselves" (Frankfurt 1997).

Affirming the singularity of human beings in no way authorizes to subject them to discrimination and unequal moral treatment. Neither does it presume that every member of society is equally prepared to deliberate and engage in the reflective equilibrium that Rawls puts at the base of justice as fairness. It will always be the case that the privileged are better equipped to see after their interests, whereas a vast majority needs tutoring and protection, at least until they become empowered to look after themselves. Human rights are expected to safeguard intrinsic vulnerability, whereas protection takes care of the specific frailties and harms that threaten and assail the human condition.

The demands of complex modern societies go beyond mere protection against physical harm and patrimonial dispossession. Sovereign power has the authority to act as arbiter and administrator of social arrangements, but this authority entails that rulers protect their constituents. Central protection has evolved beyond guaranteeing safety, being expected to cover catastrophes, epidemics, implement hygienic and sanitary measures and, in advanced societies, provide variable degrees of social security and medical care. An ethics of protection proposes to anchor security in a stable and reliable social fabric resistant against political and economical vagaries.

Protection is at the base of the State's responsibility for its citizens, and ought also to be the fundament of social institutions: in order to pursue a just social order, the weakest have to be protected so they may partake of fair arrangements. At the interpersonal level, people are only able to communicate and cooperate if they can count on mutual protection from misdeeds and abuses, resting assured that the worse-off will be sustained by the stronger, be it in childhood, impairment, disease, or senescence. State legislations consistently commit parents to take due care of their children, making neglect severely punishable. One of the most pressing problems of modern society is the lack of adequate protection for the rapidly increasing social segment of the elderly, who are no longer supported by traditional family ties or personal commitments.

Protection is also at the basis of H. Jonas' principle of responsibility, when he illustrates the primary and unsolicited protection owed to the newborn and to future generations. E. Lévinas has written movingly about the pristine ethical experience of the "I" encountering the "other" whose visage expresses a plea for protection, and initiates an ethical commitment that emerges whenever helpless need is encountered. Even utilitarian ethics based on the Scottish doctrine of empathy/sympathy, will require social morality and a legal system to protect citizens in matters that do not depend on sympathy and solidarity, like violence, poverty, or defective education (Brandt 1996).

Current sociology has devoted much thought to the loss of protection in contemporary society. Modern citizens live in "uncertainty, insecurity, and lack of protection"; as national territories and State responsibilities are demoted, individual citizens lose all mantels of social protection, having to assume personal responsibility for their own safekeeping by entering the market of insurance schemes (Bauman 1994). Protection in late modernity becomes a commodity that drives additional nails into the coffin of equality, social security, morality, solidarity, and shared responsibility. Globalization and its penchant for law and order has fueled a regression of social protection to the Hobbesian minimum, weakening the State to a mere local police station (Bauman 1998). R. Castel gives up on justice: "Modern democracy does not imply the strict equality of social conditions for it is, and will remain, strongly stratified. Rather, it demands the robustness of protections." The times of universal protection are gone, the future will at best preserve a trickle of protective policies for the most destitute; at worst, Hobbes' *homo homini lupi* might reemerge (Castel 2009).

As anticipated, an ethics of protection is radically suspicious of alleged universal equality, so often declared but never obtained. Social contracts, original positions, communities of communication, or any other model posing individuals as equally seasoned in the art of deliberation, are figments of scholarly imagination. Quite to the contrary, human societies are stratified, steeped in quests for power that invariably schemes to perpetuate itself, oblivious to justice or lack of it. Justice and inequity are conditions that evolve asymmetrically in social relations, with a penchant for loudly proclaiming equity without showing significant efforts to realize it. While the idea of justice is infused with top-down idealizations, the applied forms of ethics are spurned by injustice and the urgency to reduce it. Thus is explained that justice will not descend upon mankind nor be borne by its deeds.

Bringing protection to the forefront of ethical reflection emphasizes that the applied or practical aspects of ethics claim precedence over theory. Bioethics has fallen prey to the unrealized idea that patients and physicians, subjects and researchers, public health policies and citizens all function on the basis of universal equality. In fact, market defenders labor under the delusion that supply and demand are regulated by egalitarian processes where social security and health care can be negotiated as fairly as furniture or cars.

Public health, it is said, places a priority on the "commitment to social justice" that "attaches a special moral urgency to remediating the conditions of those whose life prospects are poor across multiple dimensions of well-being" (Powers and Faden 2006). Unfortunately, the use of indeterminate language like "social justice" and "multiple dimensions of well-being," deflects attention from the essential and uncompromising commitment to attend basic needs before scarce resources are fanned out into less essential needs and desires. Protection means covering the needs and assuaging the suffering of those afflicted by unfortunate circumstances they are not able to remedy by themselves. To protect is to honor a "principle of rejecting injury" (O'Neill 1998). The failure to protect, either from neglect or by actively denying assistance to the point of increasing harm, is such an unmistakable instance of ethical wrong-doing that it seems hardly necessary to defend the idea that protection of those who cannot protect themselves is mandatory by any ethical standard.

Power and Protection

Protection, etymologically derived from *tectum*—roof, abode—has its ethical roots in the face-to-face interpersonal encounter before being politically conceived as a civic right. The basic tenet of protection is that human affairs are steeped in injustice that deepens as the gradient between the well-off and the destitute relentlessly continues to rise. Admitting that pervasive asymmetries of power are the driving force behind socioeconomic disparities and health-related inequalities, the ethics of protection proposes that the powerful acknowledge the obligation of protecting those under their direction.

Material redistribution is not the answer to power gradients, for at best it produces a temporal and precarious symmetry that market conditions hasten to disrupt, quickly reestablishing the old order of entrenched hierarchies and economic inequities. Apart from being short-lived, tax-based redistribution has only worked reasonably well in closed communities or countries with small populations. International aid has been criticized because it provides material goods but fails to aid recipients in achieving the means for autonomous functioning. Policies based on redistribution of goods fail unless they are accompanied by empowerment programs that develop individual capabilities to pursue integration into the productive and cultural activities of society.

Belief that power might more readily be shared than wealth may appear as mere wishful thinking, but it is not unheard of. In fact, premodern despotism had given way to democracy which, for all its imperfections, is the form of government where political power is shared and alternated. Mary Wollstonecraft's writings are similarly inspired when she states that the "end of government, ought to be, to destroy... inequality by protecting the weak" (O'Neill 2007). J. Rawls addresses the same issue in his second principle of justice requiring "social and economic inequalities to be arranged so that they are both...attached to positions and offices open to all" (Rawls 1980). The basic condition of fair social order is that powers of office be equally available to all citizens, meaning that political power may not be accumulated, monopolized, or overstepped. Responsibility and accountability are social requirements designed to subject power to reasonable constraints.

An ethics of protection proposes that when or wherever people are incapable of meeting their most essential needs, social arrangements ought to offer the required protection and coverage. Such a general proposition must face at least three critical questions: (a) what is to be considered a need requiring aid? This issue has been dealt with in Chap. 2; (b) why plead for protection and not insist on care, justice, or equal opportunities? (c) How does protection avoid the shortcomings of paternalism and protectionism?

Ethics of Care

Ethics of care is a move in the same direction as protection, proposing that those with know-how and commitment be summoned to aid the less fortunate. At a first blush, the ethics of care would seem to blend greater sensitivity to need with protective attitudes, suggesting that the idea of an ethics of protection is redefining an already well-entrenched ethics based on compassion and care (Gilligan et al. 1988). Care will proximally comfort and support a slow healing process, rather than actively solving distant conflicts and averting more generalized harm. In the healing professions, there is a historical tendency to think of physicians in terms of curing agents, as distinct from nurses as caring persons, but this is of course an artificial anachronism, for therapy and care are indissoluble and synergistic. Feminism has been commended for advancing an ethics of care that is based on direct interpersonal relationships which could form the basis of a public health policy by encouraging care at the social level (Roberts and Reich 2002). A proposition of this kind can be interpreted as a preliminary idea leading to an ethics of protection.

Care is a matter of interpersonal relationships engaged in unidirectional or, more rarely, reciprocal assistance, but it also relates to the State's duty to care for its citizens. The State's welfare services of sanitary and social interventions must be complemented with personal care provided to the dependent persons. By emphasizing that people should help themselves before resorting to external aid, the doctrine intended to dignify work and personal effort, neglecting to consider the numerous individuals who live in such destitution and marginalization that they lack the capabilities to initiate the efforts required by subsidiary arrangements. All in all, care is a proximal, context-bound assistance whereas protection is a personal and political attitude of sustaining and empowering the weak.

Paternalism

A frequent reaction to the proposal of an ethics of protection is to denigrate it as a disrespectful and obsolete form of paternalistic dealing with individuals or collectives whose weakness and lack of purpose requires guidance and representation. Paternalism is broadly defined as "the interference with a person's liberty of action by reasons referring exclusively to the welfare, good, happiness, needs, interest or value of the person being coerced." In its pure form, benefits and coercion affect the same person, whereas impure paternalism restricts autonomy for the benefit of others (Dworkin 1972). In an update, the definition of paternalism includes interference with autonomy, absence of consent, and benefit for the person (Dworkin 2005). Paternalism may be justified when forceful enough reasons are presented to override someone's autonomous decision because [s]he is deemed to be irrational or excessively emotional. As for unjustified paternalism, it refers to cases where autonomy is willfully ignored and coercive pressure is applied without legitimate cause or reason. Since justification of paternalistic intervention is subject to endless controversy, it seems preferable to distinguish between authoritarian and protective paternalism.

The authoritarian form coerces and illegitimately opposes someone possessing the competence for autonomous decisions. Authoritarian paternalism is only exceptionally acceptable in emergency situations, when the exercise of autonomy might lead to harm third parties or the common weal. The more frequent form of paternalism is protective, obtaining when someone is unable to take autonomous decisions due to immaturity, defective development, loss of mental competence, or in cases where extraordinary circumstances severely weaken a person's discernment—intense suffering, fear-inspiring or otherwise stressful situations. In these cases, paternalistic protection may not only be justified, but downright mandatory, as occurs with the paternal obligation to care for children, the provision of medical assistance for individuals unable to take decisions or be adequately represented, or when imposing public health interventions in the wake of sanitary emergencies. And yet, protective paternalism is not analogous with an ethics of protection, because it lacks the clear outlines and the necessary side constraints which will be described as essential to an ethical concept of protection.

Paternalism in Public Health

Public policies tend to evoke protests and the accusation of interference with individual autonomy, usually being accepted as inevitable consequence of governmental business: citizens pay taxes, abide by multiple regulations, and obey law and order requirements. In contrast, public health measures that either lack a solid scientific basis, or may be technically adequate but little understood by the public, frequently elicit dissent and opposition. In health matters, as previously noted, autonomy and the common weal often stand in opposition: what is good for populations may be indifferent to individuals; inversely, defending autonomy may jeopardize social actions. Paternalism is a much favored issue of contention when public health indications become unpleasant and seem to interfere with personal freedom beyond what people consider reasonable and justified. For their part public health officials declare that prudential recommendations go unheeded and are of little use unless some form of pressure is imposed to obtain the necessary adherence and discipline that will bring campaigns and policies to the desired end.

Compulsory public health programs inevitably interfere with people's autonomy, implying or even disregarding their consent in the name of a beneficial purpose. Arguments have been presented to defend that public health policies are not paternalistic because the most essential aspects of "deep" autonomy are being respected; individual consent is therefore implicit, not ignored, when public policies are borne by democratic processes. Finally, solidarity and justice vouch for a beneficial overall effect even if some individuals feel mishandled (Nys 2008). In other words, the good intentions of public health programs exonerate them from being accused of paternalism. These arguments have been found wanting, for any interference with autonomy is inevitably paternalistic, even if justified (Holland 2009).

A more vigorous position that should also help differentiate paternalism from protection, asserts that public policies are by nature not paternalistic but coercive, for they do not override individual autonomy, they simply state what is accepted social ethics: people need to regulate their exercise of autonomy for the sake of avoiding harm to others, as Mill forcefully argued in his time. But coercing individual autonomy without providing adequate justification fails to mitigate the tension between personal liberty and the common weal. Of greater concern is the danger that paternalistic imposition, if not warranted and carefully monitored, might deviate in abusive directions as history, alas, illustrates too often. When public health policies are promoted without coercion, they are indulging in what libertarian paternalism understands as "nudging individual choices in welfare-enhancing directions yet without imposing any significant limit on available choices" (Ménard 2010). Nudging may satisfy libertarian ethics, but it will hardly help public health getting the job done. Ethically sustainable public policies will move along the narrow path between persuasion and coercion that is only acceptable if explicitly and convincingly justified (Faden and Beauchamp 1986).

Medical care and practice have been traditionally paternalistic, unwilling or unable to shed the critique that patients' autonomy is not being recognized and respected. Scholars have shown indefatigable zeal in addressing these issues at the clinical encounter level, though barely touching on their relevance for public health activities. And yet, medical paternalism continues to dominate in clinical practice and, what is highly problematic, extends to the relationship between researchers and subjects, especially in clinical trials where actual patients are recruited as research subjects. The "institutionalization of bioethics…does not systematically translate into more rights and empowerment of patients in everyday care" nor, it may be added, in clinical and epidemiological research settings (Orfali 2011).

Protectionism

Protectionism originally described international trade policies employing import barriers to protect the interests of countries defending local industries from foreign competition. This eminently defensive policy has been put to use in biomedical research discussions concerned with protection of vulnerable subjects from being recruited for research purposes. Exclusion argued as defensive protectionism has been decried as overprotection and paternalism that supposedly denies these individuals equal access to research of their specific needs, and discourages therapeutic studies that might lead to a better understanding of the problems plaguing such especially vulnerable groups as premature babies, the aged, the disabled, and other disadvantaged populations. This is true insofar as their problems often receive little attention because research and development are more concerned with marketable objects of inquiry. On the other hand, safeguarding weak or fragile individuals from the risks of research is intended to avoid the abusive practices of including them in studies irrelevant to their condition (Kahn et al. 1998).

The argument has been overextended to non therapeutic trials—where the study is unrelated to the needs of subjects—venting the proposition that every member of society, without exception, is under obligation to cooperate with medical research, be it for therapeutical or general knowledge seeking purposes. Accepting such a universal obligation to participate in research erodes certain basic tenets of bioethics, like autonomy, voluntariness, and the primacy of care over investigation. The recent history of biomedical research shows tragic example of disrespectful and harmful inclusion of orphanage children to study transmission of hepatitis—Willowbrook State School—hospice inmates subjected to cancer research—Jewish Chronic Disease Hospital—and, of course, the well-known Tuskegee Valley study where for decades the natural course of syphilis was observed in Afro-American population being denied available therapy.

Abusive protectionist arguments have been brought forth to waive care of helplessly suffering lives, and recruiting the incompetent for research that is vaguely hailed as aiming at protecting the common good or furthering scientific progress: "At this point in the development of biomedical science, failure to reexamine and reassess the reigning research dogma of...protection for the vulnerable...constitutes self-righteous, but culpable, moral blindness" (Rhodes 2005). Harsh words that misconstrue "a dogma of protection," thus allowing pragmatic considerations to overrun well-entrenched ethical norms. Reducing protection of individuals for the alleged benefit of science will hardly contribute to reestablish trust in the biomedical research, not all of which is innocent and devoid of vested interests unconcerned with the pursuance of knowledge. Defensive protectionism is confusing because it works both ways: defending the weak and vulnerable from undue intervention, but also serving to defend researchers and other agents from being taken to task for overstepping moral boundaries by treating subjects as mere means.

The double-edged defensive purpose of protectionism is subject to conflicting views, as well as arbitrary interpretations which fatally creep into the infelicitous issue of torture. The eventual prospect of protecting the innocent is contrived to allow torturing with the hope of gaining information that will avert major harm. This pragmatic reasoning, dubious as it may be, leads philosophers to state that "though the idea that all moral principles are absolute principles might have once been common, this is no longer true" (Lee 2007). Hence, absolute prohibition of torture becomes a matter of debate, and Amnesty International's assertion that the "use of torture is an affront to human dignity that can never be justified" is no more than a *flatus voci*.

Distortions and abuses of protectionism show a suspicious disposition to kindle philosophical dispute, confirming it to be a weak defense of negative rights that ought to be clearly distinguished from an ethics of protection, which will be true to its commitment whenever fulfilling two conditions: (a) to protect those too weak and disempowered to fend for themselves; (b) to protect in such a way as to help the frail shed their dependencies and begin to gain capabilities and empowerment.

References

Arendt, H. (1958). *The human condition*. Chicago: University of Chicago Press.

Bauman, Z. (1994). *Postmodern ethics*. Oxford Cambridge: Blackwell Publishers.

Bauman, Z. (1998). *Globalization. The human consequences*. Cambridge: Polity Press.

Beauchamp, T. L. (2004). Does ethical theory have a future in bioethics? *The Journal of Law, Medicine & Ethics, 32*(2), 209–217. 190.

Brandt, R. (1996). *Foundationalism for moral theory*. Calgary: University of Calgary.

Castel, R. (2009). *La montée des incertidues. Travail, protections, statut de l'individu*. Paris: Seuil.

Deppe, H.-U., & Regus, M. (1975). *Seminar: Medizin, Gesellschaft, Geschichte*. Frankfurt a.M.: Suhrkamp Verlag.

Dworkin, G. (1972). Paternalism. *The Monist, 56*, 645–684.

Dworkin, G. (2005). Moral paternalism. *Law and Philosophy, 24*, 305–319.

Faden, R. R., & Beauchamp, T. L. (1986). *A history and theory of informed consent*. New York, Oxford: Oxford University Press.

Farmer, P. (1999). Pathologies of power: rethinking health and human rights. *American Journal of Public Health, 89*(10), 1486–1496.

Farmer, P., & Campos, N. G. (2004). Rethinking medical ethics: A view from below. *Developing World Bioethics, 4*(1), 17–41.

Frankfurt, H. (1997). Equality and respect. *Social Research, 64*, 3–15.

Gilligan, C., Ward, J. V., Taylor, J. M. (1988). *Mapping the moral domain*. Cambridge: Harvard University Press.

Hobbes, T. (1978). *Leviathan (1651)*. Glasgow: Williams Collins Sons & Co.

Holland, S. (2009). Public health paternalism—A response to Nys. *Public Health Ethics, 2*, 285–293.

Kahn, J. P., Mastroianni, A. C., & Sugarman, J. (1998). *Beyond consent*. Oxford, New York: Oxford University Press.

Lee, S. P. (2007). *Intervention, terrorism, and torture*. Dordrecht: Springer.

MacIntyre, A. (1984). *After virtue*. Notre Dame: University of Notre Dame.

Ménard, J.-F. (2010). A 'nudge' for public health ethics: Libertarian paternalism as a framework to ethical analysis of public health intervention? *Public Health Ethics, 3*, 229–238.

Nozick, R. (1980). *Anarchy, state, and utopia*. Oxford: Basil Blackwell.

Nys, T. R. V. (2008). Paternalism in public health care. *Public Health Care, 1*, 64–72.

O'Neill, D. I. (2007). *The Burke-Wollstonecraft debate*. University Park: University of Pennsylvania.

O'Neill, O. (1998). *Towards justice and virtue*. Cambridge, New York: Cambridge University Press.

Orfali, K. (2011). The rhetoric of universality and the ethics of medical responsibility. In C. Myser (Ed.), *Bioethics around the globe* (pp. 57–75). Oxford, New York: Oxford University Press.

Powers, M., Faden, R. (2006). *Social justice*. Oxford, New York: Oxford University Press.

Rawls, J. (1980). *A theory of justice*. Cambridge: Belknap Press.

Rhodes, R. (2005). Rethinking research ethics. *The American Journal of Bioethics, 5*(1), 7–28.

Roberts, M. J., & Reich, M. R. (2002). Ethical analysis in public health. *Lancet, 359*, 1055–1059.

Chapter 6
Ethics of Protection II: Basic Outline

Abstract Protection is not an ethical principle as such, for it needs to be contextualized and adapted to a variety of situations and circumstances, at the same time requiring structure and specification to support practical decision-making. The main purpose of protection is to recognize and attend the needs of the helpless, appealing to the powerful to safeguard and secure the basic goods required for biological survival and social empowerment. Beyond providing material support for basic physical needs, protection furthers the acquisition of essential capabilities required to become integrated and cooperative community members: medical care, education, training in basic skills. Social protection includes institutions and arrangements to remedy the effects of disrupting events—disease, severe material losses, unemployment. In addition to being need-sensitive, protection is distance-neutral, acknowledging the destitution and help requirements of the marginalized and the distant.

The ethics of protection focuses on national goals, but also calls for international obligations to take care of world-wide destitution, inasmuch as the vested interests of globalization are holistic in their ecological and socioeconomic effects. Global commitment is often required as *ex post* responsibility for past colonial and exploitative politics, but protection emphasizes *ex ante* responsibility: present and future international ventures and agreements ought to include in their inception proactive and comprehensive protective commitments.

Keywords Basic goods • Biological needs • Compensation • Empowerment • Social protection

The ethics of protection partakes of M. Walzer's proposal of fragmenting justice into spheres: "the principles of justice are themselves pluralistic in form; that different social goods ought to be distributed for different reasons, in accordance with different procedures, by different agents; and that all these differences derive from different understandings of the social goods themselves—the inevitable product of

M. Kottow, *From Justice to Protection: A Proposal for Public Health Bioethics*,
SpringerBriefs in Public Health 1, DOI 10.1007/978-1-4614-2026-2_6,
© Miguel Kottow 2012

historical and cultural particularism" (Walzer 1983). Analogously, there is no solid theoretical block to create or defend a principle of protection. Quite to the contrary, protection should be the moral kit that inspires protective attitudes and actions attuned to circumstances, needs, cultural idiosyncrasies, and solutions in all the diversity that reality and the *Lebenswelt* present. Such a pluralistic view requires protection to be subject to specification and side constraints, lest it become a diffuse bona fide attitude of little practical consequence. The purpose of Chaps. 5–7 is to outline certain features that are characteristic of protection, allowing the distinction of this approach from other ethical views. By its nature, public health is the social practice where protection appears as the most appropriate ethical impulse to explain and justify its policies and actions.

The basic tenet inspiring an ethics of protection is need-sensitive. The ethics of protection intends to bring the destitute into forceful visibility and recognition, capable of triggering the social obligations to provide assistance and security. Furthermore, this ethical approach proposes to defrost authoritarian relationships and let power flow at least to the extent that all human life may become capable of sustaining itself.

The general and initial approach to an ethics of protection requires specification in order to become a plausible candidate for practical application. Protection is no substitute for justice, it rather suggests dealing with need in view that social justice is not forthcoming and seems indefinitely projected into a distant future. The role of an ethics of protection is to secure and support human beings in reaching a level of capabilities—dignified status of functionality—that will allow them to participate as equals in designing a fair social order. To this effect, protection must regulate interpersonal relations, social institutions, and public policies, establishing an ethical framework that takes care of those whose moral luck has deprived them of the means and capabilities to take care of themselves. The practical turn from the ethics of protection to a protective bioethics is aimed at countering the criticism that redundancy and a "presumption of undecidability" have driven much of bioethics into endless polemics and the refusal to assist in offering guidelines for decision-making and problem-solving (Ashcroft 2008).

Should by some unforeseen chain of events utopia become reality and human beings attain empowerment and the capabilities of covering their needs, and if social services—security, health care, education, and training—became universally accessible, then protection would turn into a supererogatory moral sideline akin to friendship or loyalty: nice to have but no longer indispensable nor ethically required. This same avowed temporality indicates how protection will differ according to external circumstances, at no point claiming to be a principle.

The Many Faces of Protection

Protection has entered only obliquely into the discussion concerning the right to health which, "like all human rights, imposes three types or levels of obligations on State Parties: *respect…protect* (prevent third parties from interfering with the enjoyment of the right to health, and *fulfill…*" (Gostin and Archer 2007, italics in the original)).

Protection is to be tailored to the actual needs of people, covering bodily deprivation, profound psychological distress, and social disempowerment. Basic needs are not analogous for sick persons requiring health care protection to those of the aged who may need support to get their daily chores done or have their bodily care attended to. In less developed countries dire need presents as unmitigated insufficiency of corporeal conditions to survive. Once this basic stage is satisfied, need becomes a matter of empowerment to gain the capabilities required when facing the challenges of social integration. Lack of empowerment in complex societies becomes a dire need, for the disempowered will not survive unless they participate in social integration and cooperative participation. Consequently, thoughts on protection lean heavily on philosophers like Sen and Nussbaum who have made empowerment the center of their attention.

Organic Needs

Centering on need is a bottom-up perspective focused on the actual deprivations people suffer. Basic needs are anchored in biological deficiencies that threaten physiological functions and survival of the organism. The weak are unable to produce the goods to satisfy their basic necessities; they lack the means and capabilities to communicate their affliction, claim basic human rights or register in programs of assistance. Due to structural violence the "poor are not only more likely to suffer, they are also more likely to have their suffering silenced" (Farmer 1996). It is a tragic paradox that the most deprived and in need of assistance, are displaced to the margins of society and beyond, becoming invisible and voiceless exiles: they are the "invisible vulnerable" (Stone 2003).

Acknowledging the destitution of the silent is therefore the first step toward an attitude of recognition that grasps and dimensions the magnitude of indigence. Instituting emergency distribution of goods may temporally assuage deficiencies but will not avoid relapses and perpetuation if not followed by the provision of elementary tools for self-subsistence. The collapse of self-sustenance—autopoiesis—of living beings that affects major portions of humanity can only be met by active protection in the form of food, shelter, and medical care at the organic level, as well as support, physical fitness and expanding capabilities for enhanced functionality.

Philosophers have pondered whether "the right to subsistence is a basic human right," coming up with a variety of pro and con arguments. Bioethics may well be excused from taking part in such discussions as would deflect it from posing a fundamental framework of decency in human relations, including the undisputable obligation of attending the basic rights of subsistence (Ashford 2009).

Empowerment Needs

A second level of need that remains hopelessly unsatisfied is the empowerment required to autonomously function in a cooperative and productive way within society. Social assistance requires knowhow transfer, technical support, and the opportunities to develop basic skills. These capabilities require recognition and respect, for they address human beings no longer as mere organisms in need of basic sustenance, but as persons engaged in actively functioning and participating in the moral vocabulary of social intercourse. As they become empowered to be moral agents, people will acquire self-respect and become worthy of the respect of others (Honneth 1997).

Social Needs

Functioning as a member of society entails a third category of needs focused on safety and security to confront unintended unfavorable circumstances and events that may cause regression to disempowerment. This cluster of needs entails assistance to overcome deficiencies that cannot be self-repaired, threatening with sliding the bearer back to destitution unless supported by social arrangements that guarantee fairness in security and health care. Programs must be devised to compensate for undeserved failures—brute luck—in addition to sustaining those not yet or no longer capable of taking care of themselves.

Responding to essential needs which human beings are incapable of covering by self-agency engages public health as a social practice that addresses health-related threats and deficiencies. At the organic level, public health protection will assist with food programs, primary medical care services, mass immunization, and basic sanitation; at the empowerment level, health risks that interfere with mastering capabilities are to be taken care of through preventive and remedial actions in the form of environmental, occupational, and individual health care and medical programs. At the social level, public health is called on to provide policies and interventions that support security in terms of unconditional access to medical assistance, removing or mitigating impairments due to chronic disease, age, or incapacitating accidents.

Distance

In addition to being need-sensitive, ethics of protection is distance-sensitive. The deprived are not only silent because too weak to solicit help they are equally invisible when too fringed out or excessively distant. Recognizing need depends on proximity, easily leading to exclusively focalizing attention in the immediate local realm. Widening the visual field to develop a more inclusive perspective requires understanding protection also as a top-down, agent-centered orientation that can be thought of as a system of concentric circles.

Dealing with collectives tends to obscure the fact that populations are aggregates of individuals, and that assuaging needs means attending to each human being in particular. Targeted at populations, public health policies finally aim to achieve proximal effects on individuals. The proximal circle, much like Walzer's "sphere of security and welfare," and analogous to care, is anchored in family and community, both being institutions based on membership, therefore setting limits of inclusion/exclusion, and creating discriminations that a decent society must compensate. Immediate protection blends into an intermediate circle of social institutions that secure education, advice, and training in basic skills and capabilities. A social order committed to fairness and security is established, where public health attends to diseases and body frailties that hamper functioning, sustaining a healthy environment to further education and occupational training in quest of unhindered development.

Shared citizenship is the common bond of this social circle of protection, and the obvious protective agent is the State and its institutions, as people interact in work, cultural associations, and religious communities. However, the recognition of citizenship entails exclusion of those who, living in the same national territory, are marginalized and exiled because they lack the legal requirements to qualify as citizens (Fassin 2006). An ethics of protection must reach out to safeguard those who are margined from sharing bonds of social and public life. The external circle of protection addresses the unemployed, the retired poor, the homeless, the uninsured sick, and those singled out by M. Nussbaum when she points out that "failure to deal with the needs of citizens with impairments and disabilities is a serious flaw in modern theories that conceive of basic political principles as the result of a contract for mutual advantage" (Nussbaum 2006). The protective role of public health at the external level takes care of disease transmission and endemic illness, the health-effects of massive migrations, crippling professional brain drain, and exploitative biomedical research practices.

Finally, distance neutrality requires acknowledging needs beyond the borders of nations, in a wider circle than the intermediate social and external national one. Entire populations are submerged in hopeless destitution, living in countries too weak or unresponsive and unwilling to take care of its people (Beyrer et al. 2007). Moral attitude toward the distant needy has been a subject of polemic discussion and half-hearted support, but of no practical consequences whatsoever. The idea of private enterprise or nations helping nations is too implausible to be pursued and would, at best, fall in the realm of altruism and supererogation, which remain outside the turf of prescriptive ethics.

As previously noted, global justice has been eroded by contextual considerations, reaffirming that each nation must come up with the social policies it can afford (Arras and Fenton 2009), and admitting that international assistance is unwilling to go beyond a minimalist program of negative duties not to harm (Daniels 2006). Many First World scholars have employed the lifeboat analogy to justify limitations of international assistance, arguing that the lavish but precarious economy of rich nations precludes the imperfect duty of aid to the distant. Unfortunately, negation is often topped by exploitation in the form of contractual arrangements that are to the disadvantage of the worse off, in fact adding insult to injury.

An ethics of protection called on to support those in need who are too disempowered to get the necessary foothold in life to develop autonomously, cannot make halt at the margins set by families, communities, or nations. Global poverty across nations and regions must also be included in a concept of protection that shuns provincialism and discrimination. An outer, global circle of protection must be envisioned even though, as argued when discussing global justice, it cannot identify a responsible agent. Nor can global protection rely on bland international agreements in which empirical evidence has shown no power or influence beyond the issuance of nicely phrased but de facto sterile commitments. The best that can be hoped for is an international order that will devise treaties that do not exert nor tolerate exploitation or coercion, fostering that the benefits of their intervention be locally shared, and, hopefully redressing past abuses. Put more bluntly, existing commercial treaties ought to reduce their exploitative clauses and be judged as to their fair treatment of weaker partners. It is unfortunate that current international agreements rarely heed the ethical requirement of protecting the weak, as crassly illustrated by trade treaties—NAFTA—coupled with progressively strict TRIPS clauses protecting pharmaceutical patents to the detriment of accessibility by less developed nations. The DOHA Declaration of 2001 which regulated accessibility to patented pharmaceutical products in situations of national medical urgency, lost much influence by the creation of more stringent TRIPS-Plus clauses that favored extending patent rights to the pharmaceutical industry (Smith et al. 2009).

Remedial work on existing international treaties and agreements ought to be complemented by including protective clauses in future negotiations and final documents should be evaluated by ad hoc ethics committees, proceeding in a similar fashion as regulations and surveillance of international research projects are handled. Respectful international arrangements rely on the good will of all involved, the more powerful accepting to relinquish excessive benefits, the weaker partners seeing to it that their politics and social institutions are purged of corruption and inefficiency. Appeals to good will may sound feeble and colorless, it nevertheless constitutes the transcendental condition of any ethical proposition at this global level of protection.

The ethics of protection should not be seen as a monolithic proposal of universal reach and unchanging qualities, considering that its realm is divided in circles—proximal, intermediate, national, and globally distant, served by different agents—individuals, family, communities, social institutions, State, international organizations. In order to preserve the fundamental character of protection in all the various circumstances, certain procedural aspects must be adhered to. A major distinction will consider protection in interpersonal relationships, illustrated in clinical bioethics and in the ethics of clinical research, as distinct from collective situations typically represented by public health, epidemiological research, and environmental ethics.

References

Arras, J. D., & Fenton, E. M. (2009). Bioethics & human rights: Access to health-related goods. *The Hastings Center Report, 39*(5), 27–38.

Ashcroft, R. (2008). Fair process and the redundancy of bioethics: a polemic. *Public Health Ethics, 1*, 3–9.

Ashford, E. (2009). In what sense is the rights to subsistence a basic right. *Journal of Social Philosophy, 40*, 488–503.

Beyrer, C., Villar, J. C., Suwanvanichkij, V., Singh, S., Baral, S. D., & Mills, E. J. (2007). Neglected diseases, civil conflicts, and the right to health. *Lancet, 370*(9587), 619–627.

Daniels, N. (2006). Equity and population health: toward a broader bioethics agenda. *The Hastings Center Report, 36*(4), 22–35.

Farmer, P. (1996). On suffering and structural violence: a view from below. *Daedalus, 125*, 261–283.

Fassin, D. (2006). La biopolitique n'est pas une politique de la vie. *Sociologie et Sociétés, 38*, 35–48.

Gostin, L. O., & Archer, R. (2007). The duty of States to assist other States in need: ethics, human rights, and international law. *Journal of Law, Medicine & Ethics, 35*(4), 526–533. 511.

Honneth, A. (1997). Recognition and moral obligation. *Social Research, 64*, 16–35.

Smith, R. D., Correa, C., et al. (2009). Trade, TRIPS, and pharmaceuticals. *Lancet, 373*(9664), 684–691.

Stone, T. H. (2003). The invisible vulnerable: the economically and educationally disadvantaged subjects of clinical research. *Journal of Law, Medicine & Ethics, 31*(1), 149–153.

Walzer, M. (1983). *Spheres of justice*. New York: Basic Books.

Chapter 7
Protective Bioethics

Abstract The ethics of protection sets the theoretical basis for an applied form of protective bioethics, which develops as an adjunct to the interpersonal realm of clinical ethics, concentrating on the values involved in the collective agenda of public health and ecology. The duality of proposing an ethical framework for both individual and collective actions is of special interest since traditional population-based public health has given way to prevention in the singular clinical encounter currently emphasizing personal health care responsibility.

At the individual level, protection is a voluntary, irreversible, continuous and open-ended commitment, to be served as long as the ward lacks autonomy and empowerment to obtain personal care. Protecting the disempowered must be restricted to areas where autonomous decision-making and empowerment are lacking, and reduce its intervention whenever and wherever protégés gain the capabilities of caring for themselves. At the collective level where public health operates, protection is required to engage in specific problems that require urgent attendance, presenting technically effective, sustainable and fairly distributed remedial solutions. Risk of unwanted side-effects should not burden specific groups or individuals, but be randomly distributed. By fulfilling these conditions, public health gains legitimacy to impose disciplined participation and reject individual claims for exemption.

Keywords Autonomy • Empowerment • Collective protection • Disempowerment • Individual protection • Paternalism

Protective bioethics is the applied form of an ethics of protection as outlined in previous chapters. Whereas protection is an ethical framework for public health, protective bioethics comes nearer to being an action-guide for actual practices in this field. The ethics of protection is a theoretical proposal that becomes operative if translated into practice as protective bioethics, following the development of bioethics as the paradigm of applied ethics in biomedical matters. Taking protective care of the other is the fundament of interpersonal ethics, in a similar way that

M. Kottow, *From Justice to Protection: A Proposal for Public Health Bioethics*,
SpringerBriefs in Public Health 1, DOI 10.1007/978-1-4614-2026-2_7,
© Miguel Kottow 2012

protection is the hallmark of social ethics. Bioethics must consider these two moments of protection, at the individual level and as part of social order. Thus, protection is primeval to personal encounter and, *mutatis mutandi*, social protection is at the base of collective togetherness.

Ethics of medical care and public health are inspired in protecting the sick: by individual ministrations in the clinical encounter, and by legitimating social health care interventions at the social level. Incorporated into health care and medical practice, protective bioethics addresses the ethical issues of clinical medicine, clinical research, and public health: medicine and public health policies inextricably share practices and ethics. One basic difference is that medicine is commissioned to act by a consulting sick person, whereas public health offers, at time imposes, population-based health care interventions. Protective bioethics must address public health issues that continue to be population based, as well as looking into individual health care that is taking over personal prevention and promotion.

The divide between public health and medical care has become eroded beyond recognition, clinical work being commissioned to include preventive care and health promotion at the individual level. By internalizing risk factors, epidemiology delivers population-based knowledge to be applied in the clinical encounter which now absorbs to a great extent the management of potential disease. On the other hand, as rising medical costs become personal burdens that are insufficiently covered by insurance plans, public health must provide State-supported medical care for the insolvent. Answering to these needs, most countries have developed more or less inclusive national health services, albeit with insufficient resources, inevitably condemned to indulge in a never ending quest to balance rationality with ethics. Shifts and transitions that affect the scope and tasks of public health call on bioethics to develop a comprehensive approach, taking into consideration that the distinction between clinical medicine and public health is in flux. As in other wakes of contemporary life, the distinction between the public and the private sphere is distorted and eroded (Touraine 1985).

As so often in its history, public health falls prey to political ideology and socioeconomic constraining determinants. Current public health approaches to the complexities of social actions are deflected to the so-called new approach where individual responsibility and self-care are promoted to take over preventive and medical care functions that in many countries had become integral part of social States. More cautiously, some suggestions are brought forth to develop a "co-production of health," a seemingly sensible subsidiary kind of proposition, though only applicable to social conditions where individuals are in possession of sufficient empowerment and freedom from need to be capable of active cooperation (Forde and Raine 2008). The rising costs of health and medicine-related goods are taxing national economies and reducing overall security and coverage. Entire populations are unable to muster the basic ingredients for self-help, and among the better off, economic crises precipitate a growing number of people into insecurity and insolvency. These are powerful arguments showing how social and health-related inequalities will increase unless public health is encouraged and allowed to adopt a comprehensive agenda of protection for all those who require it.

Deprivation and suffering are wrongly focused as a statistical dimension, for ultimately it is singular human beings that are affected, and assistance must reach the individual rather than complacently and unaccountably being programmed for populations at large. Public health acts through collectives, but its final aim must be to benefit individual members of society. Focusing on single persons as the final recipients of protection should not be confused with the downstream tendency of relating risk factors to individuals and burdening them with self-responsible care (McKinlay 1998).

Interpersonal Protection

Protection at the individual level has to fulfill five canons and conditions that emphasize its ethical characteristics and mark clear distinctions from paternalism, ethics of care, and protectionism. These features are voluntariness, irreversibility, continuity, inviolability, and retreat.

Condition of Voluntariness

This constitutes the initial ethical moment when, in the presence of actual need and suffering, the voluntary decision is made to either assist or indifferently turn away and not get involved. It is crucial to recognize the eminently autonomous character of a potential agent at the instant she perceives the need of protection invoked by a petitioner or, more generally, when becoming aware and acknowledging that someone is in need. Except in emergencies, health care professionals are free to accept or reject a patient's request to be attended—agent voluntariness.

The prospective protector faces a situation which must be resolved either by negation and retreat, or by granting hospitality to the distressed. Being freely taken, this decision is eminently ethical, for it is neither obligated by duty nor inspired in contingent utilitarian considerations. Protection—refuge, hospitality—is freely and responsibly granted: "To be what it "ought" to be, hospitality is not meant to pay a debt: graciously, it "ought" not to open itself to the guest [caller or visitor], nor "conform to duty," not even resort to the Kantian distinction "out of duty." This unconditional law of hospitality, if one can think it, shall be a law without imperative, without order, without obligation" (Dufourmantelle and Derrida 1997). The effectiveness of protection is not a matter of material metrics, its ethical excellence derives from the reasoned virtue and freely decided commitment to provide protective care.

Condition of Irreversibility

Once the decision to protect is taken, it becomes an irreversible duty to be unconditionally and irrevocably fulfilled, for this decision is a promise, a one-sided pact that binds the protector and shields the protégé. There is no reciprocity in protection for it is a one-way commitment that cannot be rescinded without becoming guilty of a major ethical transgression. The sustenance of protection is not conditioned by the meritorious behavior of the dependent nor can it be arbitrarily discontinued, for this would leave the abandoned worse off than before.

Condition of Continuity

Though protection is an ethical commitment freely engaged in, once irreversibly adopted it entails the duty of remaining unfailingly available for as long as necessary, that is, as long as the dependent is unable to assume self-care. Protection goes way beyond merely providing material sustenance, it is an ongoing and unflinching support leading the ward to gain empowerment and develop the capabilities to pursue his own existential goals. If the need for protection becomes chronic or even permanent, it may be necessary to delegate guardianship to an institution, for otherwise the protector will be unable to meet other responsibilities and uphold personal activities and aims. This situation is exemplified when permanent care for a demented elderly so exhausts the caretaker, that other duties are neglected unless the dependent is institutionalized. Care-giving social institutions are then expected to provide the support and consistency that personal protection can no longer warrant.

Condition of Constraint

This most important feature of protection serves to clearly distinguish it from paternalism. The needy, although disempowered, are not necessarily devoid of all capabilities, and should not be treated as utterly lacking in autonomy. Fragmented autonomy frequently presents in medical situations when patients become partially dependent in certain aspects of their life. They may be unable to earn their subsistence, however, remaining quite adroit at learning new skills or taking decisions in certain areas of their interest. Care of the aged often shows them to have lost certain capabilities, like the skill to work or supervise their personal finances, but maintain lucid alertness regarding medical decisions that affect their condition. Children usually command a patchy and variable array of capabilities, so that protectors and educators will be well advised to adjust their influence and allow autonomous decisions to develop and coexist with protective paternalism wherever necessary. When areas of viable autonomy are not recognized and respected, or progressive autonomous capabilities are not properly monitored, protection is robbed of its moral justification and regresses to unacceptable authoritarian paternalism.

Condition of Retreat

Protection is open-ended yet not limitless; it is constant but not invariant. Since its purpose is to empower the dependent, protection ought to be phased out as it becomes redundant and superfluous. To remain protective when no longer necessary violates burgeoning autonomy and subjects it to unjustified authoritarianism. The typical example is the parent–child relationship, where the maturing sibling progressively reaches out to exercise autonomy in quest of self-sufficiency. Should the parent not heed this development and insist in remaining overprotective, the generational gap will become a generational conflict. Physicians must learn the bitter lesson that those patients who initially may be too ill to exercise autonomy, resent being excluded from all informed consent procedures once they improve. Patronizing patients who are on their way to recovery and the insistence on medical paternalism invites discontent and malpractice accusations.

Collective Protection in Public Health

Being under the administration of governmental policies and subject to the unstable support of fiscal resources allocation, public health lacks the institutional authority to designate and autonomously carry out its programs. With dwindling budgets and weakened State support, public health can only address problems that are in clear and urgent need of intervention, that is, situations where the protective value of a program must be granted the highest priority. Nevertheless, since much preventive activity has migrated to the care of clinical practice and is in danger of being medicalized (Verweij 1999), public health must regain its pristine agenda: "A fresh approach to prevention requires a refocusing of attention from evidence relevant to individuals to evidence relevant to populations—even those in clinical settings" (Starfield et al. 2008).

Disease determination and causation has come to navigate in a sea of uncertainties and confusing complexities, making it difficult to identify and isolate specific pathogenic factors that might be auspiciously intervened. Globalizing the health issue and touting manifesto-like goals—health for all, global health—, as well as promoting well-being as a health care objective, brings forth elements that have disconcerting effects on pragmatically oriented public health professionals. Public health cannot be required to reach all-round goals like physical, psychical, and social health, nor can health policies be held responsible to make people happy.

On the other hand, public health has been substantially altered by the imposition of new tasks related to bioterrorism, often confusing military strategies with sanitary policies (Annas 2002; Rhodes 2004). The accountability of public health interventions is curtailed when scientific information in health and environmental issues is silenced or distorted for national security's sake (Milio 2004).

Triggered by uncertainties and restricted by reduced and elsewhere diverted resources, public health expertise must concentrate on problems of prime social importance that are in urgent need of attention and solution. The nearness of public health to its constituents is to be cultivated and preserved, identifying pressing problems by giving primacy to democratic participation and public accountability. For a public health action to be genuinely participative and ethically legitimate, four conditions are to be fulfilled: (a) recognition and assessment of a serious problem; (b) evaluation and availability of effective intervention; (c) random distribution of side-effects; (d) transparently justified individual discipline.

Assessing the Problem

First of all, a health-threatening problem needs to be identified as to its magnitude and probable causes, for causal knowledge influences the effectiveness of remedial solutions. Global enterprise, transnational private interests, and the offshoring of research endeavors have contributed to the baffling situation where national societies may not be aware whether they are dealing with problems specific to their reality, or being cajoled to concentrate on issues driven by the economic interests of some international enterprise. Nations plagued by endemic diseases may find themselves hosting molecular investigations—pharmacogenomics—instead of developing a much needed vaccine. Consequently, for public health to initiate a population-based program, it must be ascertained that the problem presented constitutes a severe local health burden that must be looked into.

Evaluating Effectiveness of Intervention

Problems that have been identified and given top priority are to be addressed with prompt and effective measures that will solve or substantially mitigate unwanted effects and threats. Effectiveness depends on applying the best means and methods in existence to avert and reduce impending threats. Such actions should not conform to what is locally available, as is often suggested. Less developed countries repeatedly suffer the experience of being restricted to locally accessible means that are known to be less effective, or to acquire second-rate merchandise—vaccines, medicines—because of meager budgets. Such inappropriate practices not only offer insufficient protection, they actually may harbor serious unwanted effects that place the target population at additional risk. First World nations have unwittingly provided pesticides, transgenic plant seeds, vaccines, and other products for public health actions in poor countries, even though these products had been officially banned in the country of origin due to their toxicity.

Under the banner of protection, public health programs should not be carried out unless effective measures have been secured, for even if a problem is pressing, improper interventions may be more harmful than the original situation. An unreliable vaccine will not only offer insufficient protection, it will also cause side-effects that may be considerable and irreversible.

Unbiased Distribution of Benefits and Risks

Considering that complications and unwanted side-effects are inevitable in any collectively applied program, it is mandatory that the distribution of negative consequences of public health interventions should occur at random, all participants having the same chance of benefiting as well as equal risks of suffering unwanted effects. The experience made in India is illustrating: when a massive nation-wide immunization with an oral antipolio vaccine was carried out, public health officials were unhappily surprised to detect a number of children who responded to the vaccine with a severe bout of the disease. Studies showed that this uncanny effect occurred only in children who suffered from an immunosuppressive condition, and that they could be safely immunized with a vaccine delivered by injection. Well aware that targeted complications are inadmissible, public health authorities revised their immunization programs and provided injectable vaccine for susceptible children.

Recruiting Individual Participation

Not before the three preceding conditions have been adequately fulfilled will a public health intervention gain the ethical legitimacy to carry out its protective program and to impose, if necessary, disciplinary instructions to the citizenry, thus facing the perennial opposition between the common weal and individual autonomy. Subjecting public health activities to the rigors of ethical legitimacy should go a long way toward social acceptance and transparent accountability of programs that require a certain amount of discipline and individual restrictions. Openness permits public health action to demand cooperation and institute, if necessary, sanctions for those who evade disciplined compliance or act as free-riders hoping to benefit without sharing costs and limitations.

The theoretical framework of an ethics of protection and the practical orientation of protective bioethics offer the most promising approach to reduce the tension of satisfying public needs without undue restrictions of individual autonomy. Population screening, massive vaccination, and obligatory immunization programs for children, and mandatory reporting are some of the issues that gain ethical legitimacy when an important problem is being addressed, technically efficient means of response are available, risks are reasonable and fairly distributed, and the public is called upon to cooperate in a disciplined manner to ensure successful public health actions.

References

Annas, G. J. (2002). Bioterrorism, public health, and civil liberties. *The New England Journal of Medicine, 346*(17), 1337–1342.

Dufourmantelle, A., & Derrida, J. (1997). *De l'hospitalité*. Paris: Calman-Lévi.

Forde, I., & Raine, R. (2008). Placing the individual within a social determinants approach to health inequity. *Lancet, 372*(9650), 1694–1696.

McKinlay, J. B. (1998). Paradigmatic obstacles to improving the health of populations—Implications for health policy. *Salud Pública de México, 40*(4), 369–379.

Milio, N. (2004). The dangers of "spin": Information, science, security, and welfare. *Journal of Epidemiology and Community Health, 58*(8), 631–632.

Rhodes, R. (2004). Justice in allocations for terrorism, biological warfare, and public health. In M. Boylan (Ed.), *Public health policy and ethics* (pp. 73–90). Dordrecht: Kluwer Academic Publishers.

Starfield, B., Hyde, J., Gérvas, J., & Heath, I. (2008). The concept of prevention: A good idea gone astray? *Journal of Epidemiology and Community Health, 62*(7), 580–583.

Touraine, A. (1985). An introduction to the study of social movements. *Social Research, 52,* 749–787.

Verweij, M. (1999). Medicalization as a moral problem for preventative medicine. *Bioethics, 13*(2), 89–113.

Chapter 8
Health Care Strategies

Abstract Public health needs to define its basic nomenclature. Public may refer to population, citizenry, society, inhabitants of a territory, and health is a concept that suffers from notorious ambiguity ranging from freedom of disease to a holistic state of well-being. Different meanings will strongly influence the traditional tasks of disease prevention and health promotion, codetermining whether screening programs will search for predispositions, subclinical disease manifestations, or genetic markers.

Prevention continues to be the main goal of public health policies but, ever since risk epidemiology targets research to the study of socioeconomic determinants of health, the New Public Health has internalized these risks and transformed collective preventive actions into individual responsibilities of health care and disease prevention. Whenever public health challenges are shrouded in uncertainty and prevention cannot be knowingly applied, it is replaced by a precautionary approach which is ethically less reliable because subject to the negotiations of vested interests.

Promotional campaigns often lack scientifically proven information. When public health resources are scarce, promotion should be used sparingly, for its effectiveness is difficult to assess. As for screening programs, they ought to respect the ethical requirement that only those conditions be explored that provide effective therapy or other clear advantages for people being screened.

Keywords Health definition • Health promotion • Precaution • Prevention • Screening

Defining Health

The following three chapters focus on a basic agenda for public health, each item of which has been subjected to innumerable polemics and controversies that have studded the literature for the past 20 years. If this discipline is expected to be more

M. Kottow, *From Justice to Protection: A Proposal for Public Health Bioethics*,
SpringerBriefs in Public Health 1, DOI 10.1007/978-1-4614-2026-2_8,
© Miguel Kottow 2012

than subservient to political ideologies and economic constraints, and to retain its historical place as an indispensable social practice, public health will have to aim at semantic and conceptual clarity in order to bridge the gulf between theory-laden academic efforts and pragmatically orientated decision makers. Currently, meaning and scope of public health need to be reassessed: "A general philosophy of public health, with ontological, ethical, and epistemological components, would provide a new foundation for public health decision making" (Weed 1999). The clarity of public health's core concepts has become eroded, stressing the need to redefine its range and priorities. Developing the philosophical foundations of public health may prove too far reaching for a social practice faced with major contingent challenges, but certain basic concepts need to be clarified (Nijhuis and van der Maesen 1994). As it stands, the vocabulary of public health is in flux, putting to question the meaning of public—community, population, society, citizenry, national inhabitants— as well as the concept and value implications of health—absence of disease, normal functioning, normative flexibility, well-being, happiness, flourishing (Cameron et al. 2006). Suggestions have been presented to de-emphasize health as the primary concern of public health, and develop physical safety as the central category of its policies (John 2009). Resources being finite, public health must refine its activities and pledge its competence to solve problems that are pressing and cannot be neglected. In the view of protection ethics, this means lifting those in need of basic medical care and sanitary services from the state of physiological dysfunction and disempowering diseases they are submerged in. Public health tasks in poor countries will therefore differ from those encountered in well-off populations, not forgetting that inequalities within societies will also make different demands on public health.

Public health has always been a social practice sailing under some foreign flag hoisted by international conflicts, politics, and economics. Coupled with medicine, public health shows a poor track record in reflecting upon its philosophical, or at least conceptual, foundations. Public health is also known as social medicine, population health care, and preventive medicine, each label having distinct connotations and political undertones. Of greater import is that the discipline purports to protect and promote health without a clear idea of what the meaning of health might be. Medicine has devoted unlimited but not very successful efforts at defining health and disease. The best definitions provided by Canguilhem (1950) and by Boorse (1975) refer to individual organisms and cannot be readily extrapolated to the collective dimension public health operates in. Canguilhem starts out by celebrating R. Leriche's brief definition of health as silence of the organs, further developing the idea that a healthy organism is capable of adapting its functionality to a wide range of external challenges, that is, health implies a considerable latitude of "normativity," or adaptive norm-setting capacity (Huber et al. 2011). As for Boorse, his definition of health centers on functionality in accordance with average norms of the species. However, if public health believes in probing the asymptomatic organism by setting up screening programs to capture individuals with occult predispositions that may, or may not, derive in disease, then silence of the organs is a poor and misleading indicator.

How can health promotion be furthered in the absence of a clear definition of health? How can a predisposition be a reliable indicator of future disease if undetermined external circumstances will influence its future development? Health is not a natural condition, for modernity teaches that "natural" is a matter of perspective and not a "true" state of affairs. Nor is health a "normal" circumstance, for normality is an abstraction, conceptual or statistical, which in the best of cases conforms to a social construct that is considered to be desirable. Rather than converging on a reduced dispersion of acceptable values, health is to be understood as a range of organic states that vary according to aging processes, sexual preferences, individual idiosyncrasies, social and environmental externalities, that should not lightly be labeled as deviant or pathological (Jones and Higgs 2010).

Pace WHO's definition of health "as a state of complete physical, mental and social well-being and not merely the absence of disease or infirmity," the practice of medicine and public health focuses on preventing and curing disease or, at least, mitigating incurable conditions and compensating the impairments they cause. Public health will fare better if it remains faithful to the commonsensical definition of health as the absence of disease, rather than accepting either holistic definitions, or molecular finessing.

Preventive actions are at the heart of public health, although other issues concerning population health need also to be considered such as, for example, national medical services, epidemiological research, occupational medicine, health-related ecological problems.

Prevention and Precaution

Having briefly looked at some conceptual aspects of prevention (Chap. 4), the changing strategies of preventive actions and policies need to be briefly looked into. The traditional preventive measures set up against ongoing epidemics were inspired by safety measures based on separating and isolating the victims in order to avoid affecting the healthy, as vividly depicted by Defoe in 1772 (2006), and Camus (1947). Dispositions taken to curb the ravages caused by epidemics were referred to as prophylaxis rather than prevention, in accordance with its meaning of vigilance and caution (Rosen 1993). Lacking knowledge about effective preventive measures once an epidemic had been declared, enforced isolation was the only prophylactic tool employed, basically aimed at the preservation of social order and the protection of property (Porter 2005). For a long time, preventive and sanitary public health programs have traditionally reacted to alarming increases in incidence and severity of contagious diseases, when fear and urgency justified taking measures often unproven beyond the experimental or even speculative stage. With advances in bacteriology, the institution of health care and sick nursing by F. Nightingale, and the eventual development of antibacterial drugs, prevention came to be complemented with public health measures that provided medical care for those affected by endemic and epidemic diseases. Hospices, pox-houses, and similar insulating institutions gave way to newly designed hospitals where in-patients were clinically studied and rationally treated. The sick were no longer objects of rejection and isolation, they now came to be thoroughly observed and explored, thus serving to sharpen clinical acumen and as training ground for future physicians (Foucault 1973).

Further expansion of scientific knowledge and technical knowhow allowed prevention to become highly successful in curbing such ravaging diseases as smallpox, diphtheria, and polio. Currently, population health is threatened by diseases that are often poorly understood; available preventive and therapeutic tools are based on uncertain and occasionally manipulated knowledge, while surveillance is hampered by migration and uncontrolled urban population growth. Current epidemic threats show highly erratic comportment. New infectious agents, poorly understood transmission processes, and uncertainties about the effectiveness of natural or artificially boosted immunity have led to haphazard preventive programs, gross prognostic miscalculations, and costly overreactions. As previously mentioned, in recent years the World Health Organization was willfully misinformed and directed to announce a virulent H_1N_1 pandemic, misleading a number of nations to overstock with vaccines that finally were not used. Decisions were taken in uncertainty, following a much discussed tendency to replace prevention based on solid knowledge, with the so-called principle of precaution.

As risk epidemiology and the New Public Health gained predominance over traditional public health interventions, G. Rose proposed a two-tiered approach to primary prevention. Low-risk collectives would benefit most from population-based strategies including mass screening, educational and promotional programs stressing life style adjustments, all expected to cause a diluted but substantial decrease in clinical disease because low-risk individuals were being targeted. Up to the present, these benefits have only been undisputed in the reduction of smoking habits. For their part, high-risk individuals were to be thoroughly tested and preventively treated on an open-ended long-term basis with uncertain results since successful end points could hardly be postulated. The tendency has been to privilege clinical activity by resorting to so-called surrogate end points that employ artificial diagnostic tests to label individuals as at risk. This clinically oriented prevention reduces public health activities to prevent epidemics and support a handful of life style regulating norms (Lauer 2007).

The ethical requirements of traditional prevention can only be fulfilled when effective, low-risk programs are made available to the exposed population. Mandatory vaccination can easily be imposed and accepted when secure immunity is available against a serious disease, as paradigmatically illustrated by the worldwide campaign that led to the eradication of small pox. If both the common weal and individual autonomy are to be handled with respect and dedication, preventive programs must be evaluated in ethical terms as to desired results, acceptable risks and side effects, and the fair distribution of benefits and burdens. An ethical requirement is that evaluation be knowledge-based and specific enough to enhance cooperation by being accountable and allowing public assessment of proposed interventions.

Precaution has been defined in numerous, often contradictory ways (see Chap. 4). Some propose that "the core maxim of the precautionary principle is that action should *not be taken* when there is scientific uncertainty about its potential impact" (Goldstein 2001). Sources such as the Wingspread Conference (Armstrong 1995) seem to take the opposite view by stating that when "an activity raises threats of harm to human health or the environment precautionary measures *should be taken* even if some cause and effect relationships are not fully established scientifically"

(Hughes 2006). Rather than an ethical principle, precaution appears as a strategic negotiation of interests, the stronger parties usually commanding decisions that may not be in the best interest of the common weal. Even though precaution has gained the favor of big business and the enthusiastic support of international organizations (COMEST 2005), its academic promoters uneasily insist on the ethical requirements to reduce uncertainties as much as possible, which is a roundabout way of admitting that precaution is a poor substitute for preventive programs solidly backed by scientific and technical knowledge.

Prudence suggests that whenever serious risks or harm are envisioned, precautionary action should be taken if, and only if, actual harm or impending threats can be averted, regardless of the benefits eventually lost. In other words, precaution should not be irresponsibly overused (Kriebel and Tickner 2001), and certainly not hailed as a moral panacea that accepts progress with little known consequences or, to the contrary, adopts adamant technophobic stances. Public health has been a frequent victim of precautionary enthusiasm in recommending armed defense against epidemics and pandemics. Hailing precaution, insufficiently tested technical devices and poorly understood pharmacological agents have repeatedly invaded the market, only to be recalled after severe side-effects and drug-induced deaths were detected. A prudential course is opposed by those who fear that excessive precaution will stifle scientific progress (Harris and Holm 1999), although they should bear in mind that progress is costly — climate change — and that civilization creates new health-related hazards. Furthermore, it is an undisputed experience that the benefits of innovations accrue to the privileged while risks befall the worse-off.

Health Promotion

The intertwinement of prevention and health promotion received a major impulse from the First International Conference on Health Promotion [Ottawa Charter for Health Promotion, 1986]. The Charter stresses and iterates that health promotion goes beyond preventive health care, and is to be understood as "the process of enabling people to increase control over, and to improve, their health." Health, again, is presented as "a positive concept emphasizing social and personal resources." In its introductory paragraphs, the Charter admits its focus "on the needs in industrialized countries," launching the idea that "the health sector must move increasingly in a health promotion direction, beyond its responsibility for providing clinical and curative services." Ottawa offers a phenomenal, though covert, endorsement to individual self-responsibility for health care, the supremacy of the medical and health-related market, weakening of the public sector, and governmental withdrawal from welfare and support (Wills and Douglas 2008). Promotion has gained a strong foothold throughout the developed world, with the unfortunate effect that public health has adopted it as a major strategy and deviated part of its never abundant resources to public relations strategies at the cost of more genuine and urgent tasks of health and medical care for the unprotected population.

Promotion is misleadingly presented as the liberal form of supposedly well-substantiated preventive campaigns, basking in the luxury of being free from responsibility and accountability. Whereas preventive programs cannot avoid resorting to some degree of hopefully justified coercion in order to be effective, promotion employs flashy and expensive methods to seduce, nudge, and manipulate, gaining favor by appeals to voluntariness. The uncertainties of this strategy, and the vagueness of promotional goals, are hardly comparable to effective preventive public health interventions, and yet promotion has been hailed as a "major form of primary prevention" (Holland 2007).

The contents of promotional campaigns are steeped in uncertain and insufficiently substantiated information, all the more so because they rely on formless appeals to fitness, well-being, good health, healthy life style, and similarly undefined goals. To a great extent, promotion exploits the allurement of media and electronic information networks, also relying on strongly emerging communication methods such as telemedicine and cybermedicine. By appealing to commonsense values, promotion resorts to moralizing, at times reverting to supplement its educational efforts at "criminalizing" behavior by inspiring legislation that prohibits or imposes certain conducts — seat-belts, helmets for motorcycle drivers, nonsmoking areas, sales restrictions (Cole 1995). Moralization is reasonable and a helpful adjunct to well-designed preventive measures, but gains a dubious slant when employed to decry private habits, or when arbitrarily introduced in times of economic hardships (Leichter 2003). The ultimate paradox is that promotion, being a tool much cherished by liberals, will nevertheless not hesitate to support laws imposing conducts purportedly aimed at protecting the common weal, thereby irritating conservatives who, having argued against social States, are now confronted with a regulatory State displaying authoritarian features (Brandt 1996; Carroll 2002; Epstein 2003).

Antitobacco campaigns are considered successful because they durably reduce the prevalence of smoking, but this effect may have been due to a number of concomitant measures imposing regulations and legal constraints, such as over-taxing cigarettes, creating smoke-free premises, restricting sales, rejecting smoking as a sign of moral turpitude. Their actual impact being incommensurable, promotional campaigns cannot be held accountable or subjected to a cost/benefit or cost/effectiveness evaluation. In times and places of scarcity, public health authorities should be wary of financing promotional activities at the cost of traditional, well-entrenched and demonstrably effective prevention and care programs. When public funds are limited there is no way of justifying promotional campaigns if resources are siphoned away from urgent and unattended needs.

Promotion aims at changing individual comportments and life styles. For a number of reasons there is no way of evaluating to what degree, if at all, these goals are actually reached. Life styles do change, but slowly and to a lesser degree than hoped for, making it difficult to assess whether these changes are due to insights delivered by public health campaigns, or happen to be associated with other social influences (Moreno and Bayer 1985). Promotion appeals to individual autonomy, regardless of the degree of empowerment people may have to exercise their freedom of decisions and action. Promoting life style changes are inevitably discriminating, for people need to have a surplus of empowerment and resources to opt out of risky circumstances, and take the opportunity of moving on to healthier lives. Furthermore, promotion addresses the complexity of external factors of disease but targets individuals to adapt, lacking causal evidence to make knowledgeable suggestions that, when followed, will strongly impinge on people's values and preferences.

Addressees are treated as medical end users, e-health consumers, clients and, if healthy, as potential patients. The promotional expansion of a health care market discriminates between obedient consumers and those who, being unable to purchase health-enhancing goods and services are ignored, marginalized, and unprotected (Castiel 2003). Promotion advance consumerism by suggesting the adoption of new life styles and giving preference to costly products that purportedly are more natural and healthier. Educational efforts are spent in convincing people to recognize a health care "need" and forsake current habits (Lhussier and Carr 2008). Promotion has pervasive medicalization effects: people are expected to listen to medical advice concerning their conduct and life style, they are instigated to seek prompt detection of risk factors, predisposing conditions and early stages of proto-disease; and, finally, they are led to prematurely embarking in a lifelong medical history that may, or may not, actually prevent clinically significant disease (Verweij 1999).

There are other ethical problems that promotion must deal with. "Health promotion seldom pauses even for a moment to consider that prevention, as opposed to treatment, is an ethical issue." Accepting that general medical practice does not distinguish promotion from prevention, McCormick laments that health promotion encourages "two things: (a) the screening of the healthy, that is, symptomless population; and (b) unmandated intervention in people's lives" (McCormick 1994). As noted, promotion caters to the liberal kind of new public policy that liberates central government from major public health responsibilities. Health promotion is expected to evolve from public "power-over" to individual "power with," which is a fine aspiration for developed societies where a majority, it is hoped, is capable and autonomous to carry on a power dialogue, but is less convincing for populations where social capital is low, political life scarcely participative, and economic gradients are appallingly steep.

In addition, promotion targets to discredit products marked as unhealthy, stimulating the tobacco industry, fast-food dispensers, and alcohol distillers to find enticing ways of retaining their clientele, thus unleashing a reciprocal escalation of persuasive stratagems where limited public health resources will lose out. The financial prudence of investing public funds in health promotion is subject to polemics that are fueled by ethical recriminations about the impropriety of interfering with people's private conduct. In sum, libertarian public health emphatically commending promotional actions despite insufficient evidence seems inappropriate for nations that must deal with restricted budgets to cover more basic health problems.

A different approach has been adopted in countries that offer a comprehensive national medical-care to its population, where health trainers (Canada) and public health physicians (Great Britain) function as individual health educators and promoters, often in close cooperation with communities. Although much discussed, and partaking of some problems in common with public health policies at population level, these initiatives appear to be reasonable because they are complementary and often parallel to universal access to medical care, sharing the purpose of improved health care without competing for resources. This kind of activity has prompted a classification of public health promotional efforts into *operational, political,* and *structural* (Scambler & Scambler 1995, quoted in Goraya and Scrambler 1998 italics in original). Operational promotion is incorporated in the health-care system, political promotion is within the reach of the system but requires political support and governmental action, whereas the structural level exceeds the capabilities of a public health system in view of the deeply engrained social realities involved, confirming that public health activities must remain focused on concrete problems that are amenable to solution or, at the very least, mitigation.

Health promotion is too deeply enmeshed in liberal, market-oriented processes to be a reliable ethical inspiration for public health initiatives. Justice is not served if promotion caters to social groups sufficiently prosperous to remain free of material constraints and dependencies, being able to contemplate alternative choices of life styles and habits. The protective effects of promotional campaigns, if real, address those sufficiently empowered to accept proposals for self-care, draining resources from programs that are needed to attend the deprivations suffered by the increasingly unprotected and underprivileged.

Screening

Hailed as an important form of secondary prevention, screening is employed "to identify people who will benefit from early detection of risk factors for, or the onset of, a disease" (Holland 2007). Traditional views distinguish between unselective or mass screening, and the selective form of studying groups of people that harbor or are exposed to increased pathogenic risk factors. In this perspective, the clinical exploration of individuals is called "case finding" (Gutzwiller and Jeanneret 1999), but there are those who believe that too much "has been made of the distinction between [population] *screening* and [individual] *testing*" (Fost 1992). There are, nevertheless, substantial differences that should not be lightly dismissed. Population screening, provided it has adequate sensitivity and specificity, will distinguish between positive and negative results, without weighting findings in relation to other risk elements, and rarely going beyond an exhortation to seek further exploration of individuals classified as positive. Individual screening is more of an early diagnostic procedure performed in the clinical setting, where patient and physician can analyze findings in relation to other influencing factors. Testing usually leads to active intervention either in the form of more intensive diagnostics, or of periodical check-ups and perhaps preventive treatment, the healthy testee thus becoming an early patient.

Effective screening procedures involve sophisticated equipment, comprehensive coverage, and the disposition of follow-up programs — additional confirmative diagnosis, periodical screening, and therapeutic indications. When resources are scarce, the need and effectiveness of screening programs should be carefully monitored, and deferred if resources are needed for urgent medical care (Holland 1993).

Never feeling quite comfortable with screening strategies, public health has been developing technical and ethical criteria to regulate these procedures. Ranging from 8 to over 25, these criteria are meant to benefit the population and also the tested individuals, insisting that only such programs be carried out that reveal important information about predisposing conditions, genetic carrier status, or early diagnosis allowing to institute timely treatment (Cochrane and Holland 1971). According to these criteria, the indication for screening procedures must fulfill three basic conditions: (a) Application to "important" public health problems; (b) Adequate knowledge of the natural cause of the disease to be intervened; (c) Availability of "treatment at the presymptomatic stage that should favorably influence outcome" (Edwards and Hall 1992). A major concern is that mass screening may yield false-positive cases if the procedure is not sufficiently sensitive, or false-negative results if specificity is deficient. Improving sensitivity will include borderline cases that perhaps never would evolve to frankly positive, whereas calibrating specificity may fail to identify those weakly affected. Sensitivity and specificity are inversely dependent: improving sensitivity will reduce specificity and vice versa, so that unavoidably a gray zone of erroneously classified individuals will ensue.

Wrong information may be deleterious, either by instilling unwarranted fear in false positives, or lulling into inactivity those false negatives who would have profited from early diagnosis and treatment. Being a mixed blessing, people should only be subjected to screening tests after being fully informed and having given their consent. In massive screening, where individual consent is impossible, campaigns should be preceded by ample information, and people ought to be reassured that they will not be tested surreptitiously or against their will. Most of these issues have received abundant exposure in the literature, except for two disquieting aspects: the undetermined and easily manipulated zone between healthy and abnormal conditions, and the case of screening strategies that do not entail benefits. Some authors have been sanguine about criticizing screening as potentially doing more harm than good (Mant and Fowler 1990). Others, though in principle partial to the "continuing need for good research on screening", find it necessary to warn that screening is "big business" under constant pressure from "politicians, the media and the public, as well as business interests" (Holland 2006).

If public health is operating at the national level and suffers chronic insolvency, the tendency will be to expand the range of normalcy. In liberal, market-oriented medicine, it is profitable to recruit, monitor, and treat an excessive number of individuals under the precautionary tenet that overtreatment is preferable to missed therapeutic opportunities. A new gradient is created between the wealthy who can afford precautionary treatment, and the less affluent having no access to such cautious measures. Better tests are more expensive and may paradoxically result in lower detection rates if they are not accessible to the poor population that eventually is at a higher risk (Grimes and Schulz 2002) .

Genetic and prenatal diagnosis may be a screening procedure, a test or, in some cases purely selective, as in preimplantation embryo selection. Genetic testing that reveals a recessive carrier of, say, Tay Sachs disease, may guide the decision of those affected to avoid marrying a person carrying the same recessive gene or to abstain from having off-spring burdened with a 50% risk of suffering the disease. Carriers of genes that determine late but inevitable onset of severe and lethal disease such as Huntington's chorea, cannot profit from early detection. Whether they will want to know in advance that a terrible disease will befall them is a matter that only the affected persons can decide.

An unsolved problem presents itself when discussing antenatal screening for diseases such as Down's syndrome, severe neural tube defects or, rarely, anencephaly. Since early detection has no therapeutic consequence, it will merely serve to alert the mother if abortion is legally or morally out of the question, otherwise allowing her to decide whether to interrupt the pregnancy. The ethics of abortion not being germane at this point, the double quandary is (a) whether to anticipate a diagnosis when nothing can be done about it, and (b) if abortion is an option, it would appear that screening is a procedure that does not benefit embryos which will be evicted. In this case, screening does not fulfill the ethical requirement of being of therapeutic value or of some clear benefit. It may be argued that screening with the prospect of abortion rests on a sliding slope that asks how far advanced a pregnancy may be for the mother to benefit from legitimately impeding the birth of an abnormal and unwanted child. In order not to pollute the screening problem with the abortion debate, the only way out is to deny antenatal diagnosis the status of a screening procedure, and to consider it a test people willingly submit to when weighty decisions depend on the result, in a situation analogous to in utero embryonic selection. Antenatal diagnosis is usually seen as a clear-cut pro/con dispute, when in fact it needs ample deliberation and the generous incorporation of many perspectives, avoiding "incorrect decontextualization" and hasty conclusions about the social and ethical issues involved (Lippman 1991). The point to be made for public health ethics is that the legitimacy of antenatal "screening" is not an issue per se, for it depends on the personal and social attitudes toward a possible decision to procure and abortion. In that case, antenatal screening should be considered an issue of clinical medicine and not a public health matter.

A multinational review of population genetic screening concludes that these procedures may be prematurely invading the biomedical market with an impetus that exceeds public information and ethical analysis. Commercial interests hail the benefits of forefront genetics without heeding public health's call to "proceed with caution" and indulge in "further discussion before their introduction" (Godard et al. 2003).

The guiding ethical approach to screening is protection. The aim is to detect conditions or predispositions in order to allow individuals to make decisions or seek medical care before irreversible damage occurs. In untreatable genetic conditions, the affected person may take decisions concerning marriage and reproduction, or avoid circumstances that might precipitate overt disease. In treatable, preclinical diseases, individuals with positive results may decide to follow medical indications, take preventive therapy, and pursue a "healthy" life style. If none of these protective measures are useful or available, screening will be futile and eventually harmful if it does no more than trigger anguish and despondency.

Together with promotion, screening has created a new medical watchfulness that blurs the distinction between illness and health, developing a form of "Surveillance Medicine" that "attempts to bring everyone within its network of visibility." Everyday life is overpopulated with risk factors, and illness is placed on a temporal axis, becoming a life-long process to be permanently monitored. It is unsettling that the clientele of the medical establishment is substantially expanded, benefitting health care providers and the pharmaceutical industry that thrives on supplying behavior regulating substances prescribed for the treatment of "the delicate child, the maladjusted child, the difficult child," while adults take pills for all sorts of conditions including, the WHO recently suggests, the quest of happiness (Armstrong 1995).

References

Armstrong, D. (1995). The rise of surveillance medicine. *Sociology of Health & Illness, 17,* 393–404.

Brandt, R. (1996). *Foundationalism for moral theory.* Calgary: University of Calgary.

Boorse, C. (1975). On the distinction between disease and illness. *Philosophy and Public Affairs, 5,* 49–68.

Cameron, E., Mathers, J., & Parry, J. (2006). Health and well-being? Questioning the use of health concept in public health policy and practice. *Critical Public Health, 16,* 347–354.

Canguilhem, G. (1950). *Essai sur quelques problèmes concernant le normal et le pathologique [1943].* 2d. ed. Paris: Les Belles Lettres.

Camus A. (1947). *La peste.* Paris: Librairie Gallimard.

Carroll, P. E. (2002). Medical police and the history of public health. *Medical History, 46*(4), 461–494.

Castiel, L. D. (2003). Self care and consumer health. Do we need a public health ethics? *Journal of Epidemiology and Community Health, 57*(1), 5–6.

Cochrane, A. L., & Holland, W. W. (1971). Validation of screening procedures. *British Medical Bulletin, 27*(1), 3–8.

Cole, P. (1995). The moral bases for public health interventions. *Epidemiology, 6*(1), 78–83.

COMEST. (2005). *The precautionary principle.* Paris: United Nations Educational, Scientific and Cultural Organization.

Defoe, D. (2006). *A Journal of the plague year (1772).* Project Gutemberg EBook. http://www. Gutenberg.org/files/376/376-h.htm (accessed October 2010).

Edwards, P. J., & Hall, D. M. (1992). Screening, ethics, and the law. *BMJ, 305*(6848), 267–268.

Epstein, R. A. (2003). Let the shoemaker stick to his last: A defense of the "old" public health. *Perspectives in Biology and Medicine, 46*(3 Suppl.), S138–S159.

Fost, N. (1992). Ethical implications of screening asymptomatic individuals. *The FASEB Journal, 6*(10), 2813–2817.

Foucault, M. (1973). *The birth of the clinic.* London: Tavistcock Publications Limited.

Godard, B., ten Kate, L., et al. (2003). Population genetic screening programs: Principles, techniques, practices, and policies. *European Journal of Human Genetics, 11*(Suppl. 2), S49–S87.

Goldstein, B. D. (2001). The precautionary principle also applies to public health actions. *American Journal of Public Health, 91*(9), 1358–1361.

Goraya, A., & Scrambler, G. (1998). From old to new public health: Role tensions and contradictions. *Critical Public Health, 8,* 141–151.

Grimes, D. A., & Schulz, K. F. (2002). Uses and abuses of screening tests. *Lancet, 359*(9309), 881–884.

Gutzwiller, F., & Jeanneret, O. (1999). *Sozial - und Präventivmedizin, public health*. Bern: Hans Huber.

Harris, J., & Holm, S. (1999). Precautionary principle stifles discovery. *Nature, 400*, 398–398.

Holland, S. (2007). *Public health ethics*. Cambridge Malden: Polity Press.

Holland, W. W. (1993). Screening: Reasons to be cautious. *BMJ, 306*(6887), 1222–1223.

Holland, W. (2006). Screening for disease-considerations for policy. *Euro Observer, 8*(3), 1–8.

Huber, M., Knottnerus, J. A., et al. (2011). How should we define health? *BMJ, 343*, d4163.

Hughes, J. (2006). How not to criticize the precautionary principle. *The Journal of Medicine and Philosophy, 31*(5), 447–464.

John, S. (2009). Why 'Health' is not a central category for public health. *Journal of Applied Philosophy, 26*, 129–143.

Jones, I. R., & Higgs, P. F. (2010). The natural, the normal and the normative: Contested terrains in ageing and old age. *Social Science & Medicine, 71*(8), 1513–1519.

Kriebel, D., & Tickner, J. (2001). Reenergizing public health through precaution. *American Journal of Public Health, 91*(9), 1351–1355.

Lauer, M. S. (2007). Primary prevention of atherosclerotic cardiovascular disease: The high public burden of low individual risk. *JAMA, 297*(12), 1376–1378.

Leichter, H. M. (2003). "Evil habits" and "personal choices": Assigning responsibility for health in the 20th century. *The Milbank Quarterly, 81*(4), 603–626.

Lhussier, M., & Carr, S. M. (2008). Health-related lifestyle advice: Critical insights. *Critical Public Health, 18*, 299–309.

Lippman, A. (1991). Prenatal genetic testing and screening: Constructing needs and reinforcing inequities. *American Journal of Law & Medicine, 17*(1–2), 15–50.

Mant, D., & Fowler, G. (1990). Mass screening: Theory and ethics. *BMJ, 300*(6729), 916–918.

McCormick, J. (1994). Health promotion: The ethical dimension. *Lancet, 344*(8919), 390–391.

Moreno, J. D., & Bayer, R. (1985). The limits of the ledger in public health promotion. *The Hastings Center Report, 15*(6), 37–41.

Nijhuis, H. G., & van der Maesen, L. J. (1994). The philosophical foundations of public health: An invitation to debate. *Journal of Epidemiology and Community Health, 48*(1), 1–3.

Porter, D. (2005). *Health, civilization and the State*. London, New York: Routledge.

Rosen, G. (1993). *A history of public health*. Baltimore: The Johns Hopkins University Press.

Verweij, M. (1999). Medicalization as a moral problem for preventative medicine. *Bioethics, 13*(2), 89–113.

Weed, D. L. (1999). Towards a philosophy of public health. *Journal of Epidemiology and Community Health, 53*(2), 99–104.

Wills, J., & Douglas, J. (2008). Health promotion: Still going strong? *Critical Public Health, 18*, 431–434.

Chapter 9
Public Health and Medical Care

Abstract Public health policies programs are to be indiscriminately accessible and fairly distributed. Their traditional agenda includes preventive and therapeutic health care services for those segments of the population that cannot afford private insurance and out-of-pocket medical expenses. To date, N. Daniels has offered the most cogent argument in favor of a right to medical care when disease impairs access to equal opportunities, but this liberal approach is unsuitable for populations where a right to health and medical care is essential but does not stand alone, for socioeconomic inequalities also occurs in education, social security, employment, and environmental issues. Public health must disaggregate sanitary and medical issues from the more general social agenda, taking specific care of disease prevention, health promotion and, for the less affluent, the provision of medical care. Resources for a national health service being always insufficient, economist will continue to search for appropriate rationing criteria, whereas bioethics advocates that rationing basic goods will inevitably create unwarranted exclusions of individuals and groups in dire need. Since there is no ethically legitimate way of justifying exclusion of the needy, bioethics will continue to insist that sufficient resources must be available to cover the medical requirements of all those uninsured and unable to finance them.

Keywords Equal opportunities • Medical needs • National health service • Resources • Rationing • Right to health care

It should go without saying that public health programs ought to be accessible and fairly applied to all who require them. And yet, wealth claims priorities and facilitates accessibility to the detriment of the least advantaged, which could at best fit into the concept of justice in inequity (Anderson 1999), but no doubt cannot be acceptable unless the basic needs and provision of essential goods have first been secured for the worse-off. Having placed need as the one criterion that should prevail over other factors, it would be utterly unfair to neglect the basic necessities of the needy

M. Kottow, *From Justice to Protection: A Proposal for Public Health Bioethics*, SpringerBriefs in Public Health 1, DOI 10.1007/978-1-4614-2026-2_9, © Miguel Kottow 2012

because the wealthier have hoarded scarce goods. To avoid such blatant abuses, essential goods and basic services including medicine ought not to be subject to the vagaries of the market.

The controversial issue of a national health service is a strictly political one in wealthier nations, as compared to being a preeminent matter of social ethics in Third World environments. High-income countries will harbor different political views that disagree whether people should acquire their medical coverage directly at market conditions, negotiating costs and benefits of insurance schemes or prepaid contracts, or prefer high taxation and thus support a national health service that will fairly dispense medical services. The discussion is redundant in those societies where substantial segments of the population are unable to pay for medical care, and will suffer the consequences of untreated disease unless the State provides social security including medical services. Governmental obligation to provide basic medical services to the insolvent is not a matter of political preference but of ethical rectitude.

Rich countries, like Norway, with a small population that is willing and able to pay high taxes, may well believe that "universal welfare programes might be more effective in achieving sustained alleviation of poverty," than targeting the "truly needy," but this seems a contradictory statement since universal welfare will have taken care of the truly needy (Westin 2008).

Disease prevention and health promotion at the population level are the universally accepted and uncontested tasks of public health. Disagreements arise when the issue of public medical care is approached, opening a veritable Pandora box of contentions that, for all their richness and variety, cling to two basic themes: universal health-related rights and allocation of resources to public health.

Right to Health[care]

Advocates of an undisputable universal right to health take their inspiration from the "Universal Declaration of Human Rights" [1948], a document that is, like the almost simultaneous "Code of Nüremberg" [1947], a staunch and condemning reaction to 12 years of terror and war, initiated with the 1933 proclamation of the German 1,000-year Reich. For the most part, human rights language refers to negative rights and, as far as health is concerned, it is limited to Art. 25 requiring "a standard of living adequate for the health and well- being…including…medical care." Such a formulation is hardly capable of eliciting a correlative obligation to provide "health and well-being." Eighteen years later, the U.N. adopted the "International Covenant on Civil and Political Rights" [1966], calling on "parties to take positive measures to reduce infant mortality and increase life expectancy, as well as forbidding arbitrary killings by security forces" [Art. 6]. In the same year, the "International Covenant on Economic, Social and Cultural Rights" formulated a positive right "to the highest attainable standard of physical and mental health" to be understood "not just as a right to be healthy, but as a right to control one's own health and body (including reproduction)," and stressing that "States must protect this right by ensuring that everyone within its jurisdiction has access to the underlying determinant of health… through a comprehensive system of health care." Art. 12 "requires parties to take specific steps to improve the health of their citizens…improving child health and workplace health, preventing, controlling and treating epidemic diseases, and creating conditions to ensure equal and timely access to medical services for all."

International covenant language is necessarily nonbinding and vague, since it must reconcile the variety of cultural and political orientations of participant nations. Treating countries respectfully has the paradoxical effect of leaving individuals all the more unprotected. Uncommitted language being the prerogative of international documents, the onus befalls politicians and scholars who rely on undefined propositions to either affirm or reject a right to health care, thus enacting unending polemics that defer positive action. The cautious turn from the negative rights proclaimed in the Declaration, to positive rights in the Covenants, is all that can be expected from the diplomatic atmosphere of international institutions. And yet, vaguely formulated positive rights remain ineffectual since no correlative obligations ensue to warrant their fulfillment. It remains unclear what is meant when positing a right to health, to health care or, perhaps, to medical care.

The problem facing well-established democracies is more subtle in appearance, and yet of grave consequences when segments of society or whole populations suffer the oblivion of the invisible, not being recognized as right-holders (Honneth 1997), therefore remaining unable to partake of the democratic processes from which they are margined. This is the situation of utter disempowerment that causes suffering and ill health, thus inspiring a rights-based approach legitimating participation to shake off subordination and marginalization (Yamin 2009). Participation in health matters was officially launched in the Alma-Ata International Conference on Primary Care [978], without having caused a notable impact on health-related inequities (Hall and Taylor 2003), thus confirming that the language of right claims is too weak to elicit binding correlative obligations.

Proclaiming that States have an "international obligation concerning the right to health" misses the point, for obligations are responsibilities towards citizens, not towards a Covenant, which means making "the right to health…*available…accessible…*of adequate *quality*…," and instigating "State Parties" to " *respect…protect…* and *fulfill*" corresponding obligations (Gostin and Archer 2007). To one's dismay, it appears that at "the national level, however, human rights are not ubiquitous, especially in the context of health" (Gable 2007).

Human rights and public health forge a strong alliance that, at least in theory, remains uncontested (Mann 1995). Highlighting human rights includes respect for democracy, participation, and the claim of specific positive rights, but it remains a matter of contention how comprehensive these rights are to be understood. Positions vary from a libertarian agenda restricted to safeguard personal and patrimonial security, to an extended program including a vast array of basic goods and services. Curiously enough, basic goods are not uniformly agreed upon and, for example, Rawls' initial list contained leisure but not health care. There is consensus, however, that the doctrine of human rights entails a lot more than merely proclaiming adherence to declarations and covenants, democracies having the obligation to address positive rights beyond the basic ones of freedom and unassailability. Unfortunately, there is also ample evidence that nondemocratic governments violate human rights in a cascade of oppression that goes from torture to denial of basic support, as has been constantly and repeatedly denounced.

Health is a hopelessly polysemous concept, subject to psychological and cultural interpretations that will not fit into the suggestive language of rights nor the binding one of obligations. Substituting health care for health is no improvement, for it still remains undefined who and what is being cared for, and by whom. Claiming that "it has been determined by nature, natural law, and natural rights that human beings have the right, not the privilege, to health care access" appears to be begging the question (Papadimos 2007). Rights do not float unbound in social space for they are always directed at a good. Right claims related to health seek tools and procedures to give substance to their demands. Thus, a right to health is presented as a right to equality, to participation, to empowerment; but again, sweeping declarations are incapable of inspiring effective action.

States invariably create public health institutions and, as the Covenants suggest, their purpose is to improve health and "ensure equal and timely medical care for all," which entails universal national medical services, and raises the question whether public health ought to honor at least, but no more than, a restricted program of preventive sanitary concerns, or take up the task of actively securing comprehensive medical hospitality. This is, first of all, an ethical question that should avoid becoming encased in endless deliberation, rather seeking to get "into the language of public policy, where it can be used to formulate policies, organize systems and services, and develop actions that realize health" (De Negri 2008).

Perhaps the most convincing school of thought in favor of a public health commitment to the provision of medical care, was developed by N. Daniels and his concept that health care is to be provided for all those who suffer from disease and therefore lose out in the turf of equal opportunities: "The basic idea is that health is the absence of disease, and diseases are *deviations from the natural functional organization of a typical member of the species*" (Daniels 1996, italics in original). His proposal has not managed to negotiate the gap between theory and practice, due to at least three objections that might be raised. First, Daniels definition of disease, based on a physiological approach (Boorse 1975), has more conceptual force than medical credence. In his view, disease means abnormal functioning as compared to standard performance of the species, that is, statistical norm defines health and disease, a concept resisted by philosophers of medicine (Goldstein 1995). Second, equal opportunity is an elusive concept, for people may have similar capabilities and yet differ in the opportunities available or attractive to them. Opportunities are context-bound, and richer societies will have a broader range than less developed nations; there are substantial differences between job opportunities and, say, opportunities of choice where and how to spend free time. Efforts at defending health care equity based on the fact that diseases must be treated to restore equality of opportunity have been found wanting for lack of consistency and definition (Sachs 2010). Recovering normal bodily functioning is essential but not sufficient to enjoy equal opportunities if people lack education, basic capabilities, certain moral attributes, and social skills to ensure empowerment. Third, should every person be entitled to receive medical treatment for diseases that handicaps her attainment of opportunities, a governmental institution would need to provide or guarantee the provision of unrestricted access to comprehensive medical care, and to institute compensatory mechanisms if therapy proved unsuccessful in restoring the state of equal opportunities.

Equality of opportunities is an individualistic approach, perhaps reasonable in societies where disease is the main reason for people to lose out in the competitive opportunity race. Daniels' theory seems tailored for societies where, all other circumstances being fairly distributed, health becomes a major factor governing life prospects. But then, egalitarian society, inasmuch as they exist, will have compensatory mechanisms to render disease less consequential: adequate medical coverage, unemployment compensation, appropriate security for the disabled and the aged. In nations that harbor severe socioeconomic inequalities, State-supported medical care assumes an obligation to remove unhealthy and disease-related conditions in order to help the needy pursue and make use of other basic goods. Health care is part of a cohort of social services, but unless health is attended to, people cannot embark in acquiring educational, occupational, and other important capabilities.

Basic medical needs cannot be prioritized, for there is no cogent argument to ration essential goods and exclude some needy recipients to the benefit of others. Consequently, deliberation and participation at this level seem out of place, for an ethics of protection will demand the universal obligation to meet *all* basic health-related needs, regardless of common opinion or political views. Participation does become important at a later stage when medical care can be programmed beyond basics, when health metrics, disease burdens, and ethical considerations guide policies to legitimately and reasonably apportion limited resources. This reasoned approach is subject to circumstantial side-constraints which make it unavoidable to embark on ethical considerations about the best way to combine fairness and contingencies. Rationing is always a second best strategy that should not substitute for comprehensive coverage of basic needs.

Allocation of Scarce Medical Resources

In the final analysis, strategies to improve access to health and medical care, whether based on rights, participation, or empowerment, are subject to the allocation of earmarked resources, a terrain where politicians, advised by economists, carry the decisive word. Decisions on resources allocation have traditionally been downstreamed, the high political echelons balancing the needs and requests of governmental institutions, subject to political interests and lobby-influences, where ethical reasoning is neither possible nor much appreciated, and participation is reduced to a mere formality. Asking "who should do what" has no ultimate answer, the debate about priorities cannot be resolved, the only rewarding attitude being to keep up a continuous, collective, open argument (Klein 1993). This sensible conclusion is, nevertheless, profoundly unsettling for two reasons: allocation of health resources engages the three Es—economists, ethicist, and epidemiologists—where ethics regularly draws the short straw, and is trumped by economists who argue that to be ethical means to be rational, and rationality implies accepting a limited budget and making the best of it. The suggestion of a deliberative fair process or "accountability for reasonableness" aims to reach a just allocation acknowledging resources constraints, stakeholders participation, and governmental accountability, hoping to justify "how to select winners from losers and how to determine which claimants should have priority" (Gruskin and Daniels 2008). The procedure appears reasonable for medical systems that have already given fair coverage to basic needs, but it would certainly be unethical to ration if losers be bereft of essential goods and services. A second major source of unhappiness with Klein's proposal of open-ended high-level discussions on available resources lies in the tendency to destabilize policies and budgets because priorities shift, and items privileged at one time are displaced by elaborate arguments or changing circumstances, in which case medical coverage becomes unpredictable.

Downstream economic policies are based on the ethical fallacy that budget restrictions are unavoidable and adequately resolved by economists and political authorities, allocations thus remaining invariable though requirements and operational costs increase. Politicians will more readily approve major defense expenses than allocate additional monies to social services like education or public health. Insufficient resources become a fiscal constant all the more restricted as the costs of medical services rise steeply. The answer is cost containment which means reducing coverage and safety. National health services enter a permanent battle to determine priorities in the assignation of resources, inspired by the premise that "scarcity is the mother of allocation." Establishing criteria for allocation is expected to be value-free and therefore amenable to economic metrics, but cannot ignore issues steeped in ethical evaluation: health, disease, and medical care. Nothing would be more unethical and dangerous than handling the human organism as a value-neutral machinery, in ways despotic regimes have exercised beyond the limits of basic humanity (Persad et al. 2009).

Alas, all rationing criteria are flawed because they perpetuate themselves; once poor prognosis or age is employed to reject certain candidates, the excluded will never come to benefit from scarce resources unless criteria are changed (Stein 2002). Even if grass-root participation is sought, minorities will remain systematically unattended. Lottery has been proposed as the only fair allocation criterion, but its insensitivity to specific contexts and situations will result in morally unacceptable discriminations as exclusion hits some cases harder than others. Even though allocation committees will have to indulge in prioritization when confronted with insufficient funds, they must keep in mind that their participation underwrites unfairness and should not deflect them from the unrelenting demand to increase the allotment of necessary resources to meet basic needs that cannot be postponed.

Health care planners deem it impossible to balance the allocation of resources in compliance with the three major constraints involved: availability of resources, needs to be met, quality and standards of care. Accepting that all three criteria cannot be fully met, economists are in constant search for the most objective way of conforming, well aware that no formula will be satisfactory: "Objectivity is a valuable property in its proper place, but it is an operational contrivance rather than an over-riding principle and must not be allowed to bar the introductions of value judgments—in *their* proper place" (Knox 1978). Being in no way insensitive to this plight, bioethics must nevertheless insist on its own perspective, by countering with three *caveats*: (1) The amount of resources devoted to health and medical care rests on political decision that needs to be in permanent revision and subject to the renewed will to increase health monies at the cost of other, expendable budget items; (2) Needs can only be considered variable and context-bound after the basic necessities of all have been met. Biological needs and basic empowerment to ably function in society ought to be secured before priorities are set for less essential necessities; (3) Quality and standards of care cannot be a matter of negotiation at the basic level. Doing less will not remove essential needs, it will only change their appearance: hunger may give way to undernourishment, insufficiently treated diseases become chronic and disabling. Bioethics asserts the right to participate and occupy its proper place in decisions on resources allocation, further claiming that its position should be substantially more central and decisive than has been traditionally the case. Installing ethical frameworks in the midst of health and medical care programs goes hand in hand with active and transparent public accountability (Newdick 2005).

The understanding that social, economic, and environmental factors influence health and disease is more prevalent than ever, causing a mounting divergence between healthy public policies as deployed by the Ottawa Charter (1986), and undisputed data showing that material inequality, health gradients, and inaccessibility to health care are world-wide on the rise (Ooms and Hammonds 2010). The holistic view of health-related external factors, being undisputed by public health, places remedial action way beyond what national efforts can envision, let alone implement. Debates in a matter charged with uncertainties and poorly substantiated opinions are made almost worthless because disputants rarely specify what they are arguing about, thus disorienting planners, evaluators, and critics, and rendering them incapable of envisioning meaningful improvements or reforms. Under these circumstances, a rational debate on rights and duties in health and medicine remains too theoretical and unspecified to be of any political import.

Priority of Basic Medical Needs

Adopting the perspective of protective ethics, the one goal that all should agree upon is to cover essential health needs, providing sufficient resources to attend these needs before embarking in other less urgent programs aimed at health enhancement or well-being. The upstream measure of adequacy is the removal of dire health deprivations, which will reflect in decreased child mortality, less severe morbidity, increased longevity and, eventually, a reduction in health care demands as people become less sick and receive timely medical care.

In order to secure effective protection, health-related issues must be downsized from global dimensions, and disaggregated from socioeconomic determinants, lest public health take refuge in the unassailability of these determinants and abide by what the new public health is promoting: since external risk factors are solidly anchored in reality, the easiest way out for politics is to insist on individual self-care.

Going against the grain of many highly esteemed views, public health must seek effectiveness by disaggregating health policies from the complex reality of social injustice and poverty. R. Klein and the Center for the Analysis of Social Policy at Bath University which he chaired, have consistently defended that the "case for greater social and economic concern should be argued on its own terms, as intrinsically desirable, rather than using the health issue to justify it" (Klein 1991 quoted in (Day 1996). Public health engaged in "maximizing population's health" by attending well-identified health care needs, should go a long way to provide protection which, in turn, will improve empowerment and make participation possible in order to reduce health and social gradients.

Meeting essential health needs is a vital part of empowerment that coincides with Sen's proposal to provide the needy with the capabilities to become cooperative members of society and exercise the autonomy of pursuing their own life-project. There is no way that a liberal market economy could, or would be willing to, fulfill such a program, which necessarily must be understood as a State policy, remaining a top priority of public health agendas of nations and social segments that live in deprivation or severe limitation of basic medical services.

References

Anderson, E. (1999). What is the point of equality? *Ethics, 64*, 287–337.

Boorse, C. (1975). On the distinction between disease and illness. *Philosophy of Public Affairs, 5*, 49–68.

Daniels, N. (1996). *Justice and justification*. Cambridge: Cambridge University Press.

Day, P. (1996). Some policy analysis fallacies: A nit-picker's guide. In P. Day, D. M. Fox, R. Maxwell, & E. Scrivens (Eds.), *The State, politics and health: Essays for Rudolf Klein* (pp. 187–205). Oxford: Basil Blackwell Ltd.

De Negri Filho, A. (2008). A human rights approach to quality of life and health: Applications to public health programming. *Health and Human Rights, 10*, 93–101.

Gable, L. (2007). The proliferation of human rights in global health governance. *The Journal of Law, Medicine & Ethics, 35*(4), 534–544. 511.

Goldstein, K. (1995). *The organism*. New York: Zone Books.

Gostin, L. O., & Archer, R. (2007). The duty of States to assist other States in need: Ethics, human rights, and international law. *The Journal of Law, Medicine & Ethics, 35*(4), 526–533. 511.

Gruskin, S., & Daniels, N. (2008). Process is the point: Justice and human rights: Priority setting and fair deliberative process. *American Journal of Public Health, 98*(9), 1573–1577.

Hall, J. J., & Taylor, R. (2003). Health for all beyond 2000: The demise of the Alma-Ata Declaration and primary health care in developing countries. *The Medical Journal of Australia, 178*(1), 17–20.

Honneth, A. (1997). Recognition and moral obligation. *Social Research, 64*, 16–35.

Klein, R. (1993). Dimensions of rationing: Who should do what? *BMJ, 307*(6899), 309–311.

Knox, E. G. (1978). Principles of allocation of health care resources. *Journal of Epidemiology and Community Health, 32*(1), 3–9.

Mann, J. (1995). Human rights and the new public health. *Health and Human Rights, 1*(3), 229–233.

Newdick, C. (2005). Accountability for rationing—Theory into practice. *The Journal of Law, Medicine & Ethics, 33*(4), 660–668.

Ooms, G., & Hammonds, R. (2010). Taking up Daniels' challenge: The case for global health justice. *Health and Human Rights, 12*(1), 29–46.

Ottawa Charter for Health Promotion (1986). *Health Promotion International, 1*(4), 405–409.

Papadimos, T. J. (2007). Healthcare access as a right, not a privilege: A construct of Western thought. *Philosophy, Ethics, and Humanities in Medicine, 2*, 2.

Persad, G., Wertheimer, A., & Emanuel, E. (2009). Principles for allocation of scarce medical interventions. *Lancet, 373*(9661), 423–431.

Sachs, B. (2010). Lingering problems of currency and scope in Daniels's argument for a societal obligation to meet health needs. *The Journal of Medicine and Philosophy, 35*(4), 402–414.

Stein, M. S. (2002). The distribution of life-saving medical resources: Equality, life expectancy, and choice behind the veil. *Social Philosophy & Policy, 19*, 212–245.

Westin, S. (2008). Welfare for all—Or only for the needy? *Lancet, 372*(9650), 1609–1610.

Yamin, A. E. (2009). Suffering and powerlessness: The significance of promoting participation in rights-based approaches to health. *Health and Human Rights, 11*, 5–22.

Chapter 10
Ethics and Epidemiology

Abstract Although epidemiology purports to be the morally neutral scientific arm of public health, it is obvious that bioethics must be involved in assessing the social relevance of epidemiological studies. Too much research funding goes to the study of diseases that constitute a less severe disease burden than the neglected diseases that are endemic in poorer countries—90:10 gap—with deprived populations plagued by high infant mortality rates, chronic weakness, and life expectancy that is less than half what members of affluent societies enjoy. Furthermore, epidemiological research must abide by the already well-established ethical surveillance applied to clinical trials, in addition to special concerns due to population research, like data-base confidentiality, postresearch benefits to host communities, safeguards against commercial use of population genomics.

Ethical concerns also apply to the intricate translational processes that intermediate between basic epidemiological research and the final issuance of public health policies and interventions. Such complex interactions may fall under the influence of experts, interest lobbyists and vote-mongering politicians, leading to top-down decisions that disregard the upstream opinion of a citizenry that is being deprived of its right to deliberate and participate in public policy matters of social concern.

Keywords Epidemiological research • Neglected diseases • Research ethic • Scientific relevance • Translational epidemiology

Epidemiology employs quantitative methods to study diseases in human populations. It began in the nineteenth century as a statistical discipline that delivered data about the relationship between poverty, urban living conditions, disease, and early death. Nascent positivism focused on studying diseases rather than social conditions, giving rise to clinical epidemiology presently complemented, perhaps even superseded, by an avid interest in molecular epidemiology. Epidemiological sociology explored and identified social factors that influence biological processes, coming up with the unsettling conclusion that useful knowledge and technology are unevenly

M. Kottow, *From Justice to Protection: A Proposal for Public Health Bioethics*, 93
SpringerBriefs in Public Health 1, DOI 10.1007/978-1-4614-2026-2_10,
© Miguel Kottow 2012

distributed, preferably reaching the well-off who also have the means of applying new insights to improve their health care. Social gradients influence the readiness to assimilate epidemiological findings—smoking causes cancer—and to act accordingly (Link 2008). In all its variants, epidemiology is clearly never the neutral fact-gathering pursuit that its practitioners and advocates claim (Porter 2005).

Epidemiology as Science

The ethics of epidemiological research has received little attention as compared to the massive academic production devoted to the ethics of clinical trials. Epidemiology has become the scientific arm of public health, providing hard data and structured knowledge that informs policies and programs. Public health researchers sway between seeing themselves as scientists dispassionately pursuing knowledge, and accepting that research objectives cannot be divorced from everyday life posing problems and seeking solutions. Max Weber, who is hailed as the champion of value-free science, in fact claimed that only the scientific method itself ought to be free of bias and ideology, admitting that it is society which values and poses the issues it expect science to tackle, finally deciding on the pros and cons of applying the knowledge gained. Most epidemiological studies take pride in being strictly positivistic, unwilling to jeopardize their allegiance to the scientific method and claiming immunity from other perspectives. The International Epidemiological Association stands on record for sponsoring an international conference where it was claimed that epidemiology is a basic medical science that needs no justification from the outside, epidemiologists carrying social responsibility as members of a community, but not *qua* scientists (Anonymous 1999, quoted in Weed and McKeown 2003).

The Ethical Issues

The first ethical approaches dealt with the professionalism of epidemiology investigators as discussed in symposia (Fayerweather et al. 1991), and documented in codes structuring an ethics *of* epidemiology, such as The International Guidelines for Ethical Review of Epidemiology [CIOMS 1999], and the American College of Epidemiology Ethics Guidelines for Epidemiologist [2000]. With the subsequent publication of general frameworks, there slowly emerges an ethics *in* epidemiology from which specific issues and areas of concern slowly branch out (Kass 2001; Callahan and Jennings 2002; Childress et al. 2002; Kass 2004).

Bioethics in epidemiology concerns the ethics of research, and the translation of scientific information to practical action and to public health policies. The question arises whether epidemiology is a purely cognitive discipline or should embrace advocacy in favor of therapeutic social intervention aimed at solving or at least mitigating the problems detected. Ethical reasoning in epidemiology gains special importance because whatever public health does or fails to do affects many people, even society as a whole, thus confirming that schematic, principle-based, interpersonal clinical bioethics is ill suited to discuss population issues.

Many scientists believe that the oversight practiced by clinical research ethics committees should not apply to epidemiology, where the risks of physical and emotional harm do not occur. Nevertheless, stringent ethical oversight ought not to be relaxed because subjects are not actually harmed but, at the most, wronged by eventually suffering social disadvantages and discrimination (Capron 1991). Some will argue that being wronged may not be harmful, thereby dismissing a basic tenet that the severity of harm ought to be evaluated by the injured, not by outside observers and much less by the harm provoking agent.

For all its surveillance by Institutional Review Boards and other sorts of research ethics committees, biomedical investigation is still plagued by disrespectful treatment of research subjects. Major transgressions are occasionally reported in specialized journals or even make the headlines, especially if the death of young subjects is involved (Gelsinger and Shamoo 2008), casting a veil of uncertainty and suspicion on biomedical research activities. Shunning the limelight of academic and social criticism, research promoters and investigators steadily reinforce the tendency to offshore projects in search of willing subjects in underdeveloped countries where ethical concerns are said to be less stringent (Petryna 2007). Unwholesome research practices have employed strategies that validate placebos, deny treatments, even going to the extremes of inducing diseases in disadvantaged subjects (Macklin 2004; Childress et al. 2005).

Scholars writing on research ethics often limit their concern to the prevention of major risks and unwanted side-effects, an attitude reminiscent of protectionism aimed at avoiding severe harm, but devoid of positive efforts to benefit research subjects by providing comprehensive ancillary care and posttrial benefits as requested by the Declaration of Helsinki. In spite of registering potentially dangerous practices like pretrial wash-out periods, sham surgeries, suspension of routinely needed medication, and denial of posttrial benefits, analyst remain mute on the ethical implications of these issues. Long forgotten are warnings that the sick, when becoming subject-patients, should only be put at risk if the trial is meant to add knowledge and improve treatment of their specific disease—therapeutic research. Similarly, epidemiologists ought not to inconvenience or burden populations, nor submit them to any risks, unless their research explores a pressing problem the community suffers and needs to alleviate.

Risks and harm in epidemiological studies differ in nature from those of clinical trials. While clinical studies often exclude benefits and may give priority to scientific excellence over social relevance, epidemiology dealing with health and disease at population level cannot freely chose its research themes, for it is under obligation to address those aspects of social reality that cause suffering, and search for ways to enhance health or even well-being. Epidemiologists stand under the ethical mandate not only to avoid causing harm with their research, but also to benefit the population under study, at the very least aiming at a decrease of the disease burdens it suffers. Research ethics is a prime example where focus on autonomy, dignity, and informed consent are subject to endless debates that neglect to directly address the prime ethical requirement of protecting subjects from undue harm.

Epidemiological research is equally plagued by tendencies to subordinate the protection of populations to other interests. This quandary results from the unending strife between research strategists who will randomize control studies using the locally available medical means as control group, and the ethical requirement of employing "the best current prophylactic, diagnostic, and therapeutic methods" in existence, as the Declaration of Helsinki and ethicist opposed to the use of placebos will have it (Angell 1997). Allowing availability criteria to prevail over excellence tempts researchers to disregard the medical needs of the subjects they recruit.

Collective studies, as well as the analysis of existing data-bases, usually preclude or at least discourage researchers from obtaining individual or communal consent, an obstacle that is compounded by the possibility of eroding confidentiality and privacy when data-bases are explored to cull and publish information that had been originally collected for other purposes. Beyond applying stringent control to guard anonymity, research done when informed consent is impossible to obtain must be justified by its social importance. These unavoidable ethical frailties make it all the more essential that epidemiologists concentrate their efforts on actual population health problems rather than catering to vested interests, academic trends, or personal inclinations. The importance of attending topics directed at health care and disease prevention imposes a research ethics aimed at the needs of populations in order to substantiate protective public health interventions. Additional criteria to ensure the pertinence of epidemiological studies include plausibility—evidence supported by reasonableness—and adequacy—demonstration that intervention has been effective (Victora et al. 2004).

The Question of Relevance

As the causes of diseases and the obstacles to attain health become more complex, eluding straightforward deterministic explanations, research also turns more intricate, forcing epidemiology to probe multileveled socioeconomic realities where questions of gradients and relative deprivation are posed, threatening to neglect the pressing but as yet insufficiently explored aspects of absolute deprivation. Stress and susceptibility to disease caused by unsatisfactory social status, feeling of inferiority and relative deprivation in social life are no doubt worthy of attention; thorough research has led to the conclusion that in developed nations individual, constitutional, and behavioral characteristics weight more heavily as health aggressors than social determinants and material conditions (Wilkinson 1994). Such conclusion, important as they may be, obscure the fact that even the most developed countries harbor substantial pockets of poverty and deprivation, epidemiologists having often noted that high national income may correlate with poor demographic health metrics.

Relevance of epidemiological studies differs in wealthy countries as compared to the persistent problems of poor environments, making it equally important to study the needs of the absolutely deprived as well as the pathogenic effects of relative deprivation. Gradient epidemiology is a key conceptual perspective, but it should neither conceal nor neglect that lack of resources continues to be a major cause of misery and disempowerment (Baker 1995).

When resources marked for research are scarce, ethical oversight ought to finesse a distinction between pertinent and relevant projects. Epidemiologists are expected to focus on actual problems—pertinence—but with increasing scarcity of investigators, equipment, or monies for running costs, national research policies will have to list priorities and apply criteria of relevance, that is, attending the most pressing pertinent problems first. Otherwise, research practices will sustain the 90:10 gap— most research funds and efforts concentrate on low-prevalence diseases—and negatively seal the destiny of "neglected diseases," those that massively affect poor population but receive low research priority. Research and marketing of new pharmaceutical agents is substantially more active in developing therapeutic drugs to the detriment of preventive products, and efforts are preferably targeted at noncommunicable diseases rather than addressing the substantially greater burden of endemic infectious and parasitic conditions. Clearly, medical problems in underdeveloped countries attract less pharmaceutical activity, with a relative neglect of vaccine development and local therapeutic needs (Catala-Lopez et al. 2010). Efforts at modifying the current "patent-based R&D system" of pharmaceutical research suggest "changing the current market paradigm by separating innovation from production and distribution, creating essentially two distinct markets" (Winters 2006).

Bioethicists have occasionally referred to applied ethics as a "component" of public health ethics, epidemiologists only recently discussing and accepting the social responsibility of being accountable to society, committed to participate in public health interventions in a reliable way, and engaged in "thoughtful public advocacy"(Weed 1994). Virtue ethics has been proposed as the core value of epidemiologists acting as scientists and as committed public servants at the community level (Weed and McKeown 1998). Beyond this general tenet, epidemiology, no less than clinical research, must abide by the requirements of ethical evaluation and submit to more energetic surveillance than hitherto practiced.

Translational Epidemiology

Knowledge that aims beyond self-referent science must find ways of translating evidence-based information into decision-making and action, allowing research results to flow into policy and practice (de Leeuw 2008). From the vantage point of ethics, one of the most interesting models relating knowledge and action is provided by B. Latour's Actor-Network Theory posing that science and its application constitute a permanent interaction between researchers and the public in the production and assessment of a socially relevant process (Krarup and Blok 2011); a similar idea is developed under the denomination "hybrid forums" (Callon et al. 2001).

The translation of epidemiological knowledge into meaningful public health intervention is especially arduous when aiming to influence individual behavior and lifestyle, reaching a maximum of "translational anxiety" in molecular epidemiology with its enormous cognitive discontinuity between basic genetic discoveries and potential applications to health enhancement, risk reduction, and disease prevention.

Basic research is accumulating information that does not deliver commensurate gains in therapy, diagnosis, or prevention (Butler 2008). The proliferation of data accumulated by molecular biology will possibly have only slight effects on current epidemiological paradigms (Winkelstein 1996). The majority of molecular studies do not pass the litmus test of relevance, which also must be applied to the multiple and intricate boundaries between translational interfaces that condemn much ground research to be short-lived and irrelevant. Although it is undisputed that knowledge improves public health's efficacy and efficiency, it is equally true that useful knowledge has become a merchandise not freely available to all who might need it, resorting as it does to legal safeguards, "such as patents, copyrights, trademarks, and confidentiality agreements" (Landry et al. 2006). Entrepreneurial interests are strong and not rarely contrary to the common weal, a fact that should be kept in mind every time the research establishment purports to be advancing science in the name of progress and the common social good.

Translation epidemiology studies the transference of knowledge from basic science $[T_0]$ to the evaluation of population-level health impact interventions $[T_4]$ (Khoury et al. 2010). The translation of knowledge in, and by means of, epidemiology is a long-winding, nonlinear process, subject to numerous financial, political, legal, promotional, and other forces. The complex transit from "bench to bedside" — in epidemiology from lab to community — is compounded by multidisciplinary influences that come to bear when applications are imminent, creating a turmoil of values, underdeterminations, and uncertainties.

Ethical evaluation remains alert to the relevance of scientific knowledge and the question of legitimate interventions amidst doubts and unreliable information. Wariness and caution are essential when evidence-based knowledge fails because its hasty implementation is "untested, unsuitable, or incomplete" (Madon et al. 2007). Being ultimately concerned with health-related protection of human beings, bioethics cannot but persevere in practicing prudent and strict surveillance over the scientific and practical aspects of public health.

References

Angell, M. (1997). The ethics of clinical research in the Third World. *The New England Journal of Medicine, 337*(12), 847–849.

Baker, D. (1995). Poverty and disease: A postcard from the edge. *Journal of the Royal Society of Medicine, 88*(3), 127–129.

Butler, D. (2008). Crossing the valley of death. *Nature, 453*, 840–842.

Callahan, D., & Jennings, B. (2002). Ethics and public health: Forging a strong relationship. *American Journal of Public Health, 92*(2), 169–176.

Callon, M., Lascoumes, P., & Barthe, Y. (2001). *Agir dans un monde incertain*. Paris: Éditions du Seuil.

Capron, A. M. (1991). Protection of research subjects: Do special rules apply in epidemiology? *Law, Medicine & Health Care, 19*(3–4), 184–190.

Catala-Lopez, F., Garcia-Altes, A., et al. (2010). Does the development of new medicinal products in the European Union address global and regional health concerns? *Population Health Metrics, 8*, 34.

Childress, J. F., Faden, R. R., et al. (2002). Public health ethics: Mapping the terrain. *The Journal of Law, Medicine & Ethics, 30*(2), 170–178.

Childress, J. F., Meslin, E. M., et al. (2005). *Belmont revisited.* Washington: Georgetown University Press.

de Leeuw, E., McNess, A., et al. (2008). Theoretical reflextions on the nexus between research, policy and practice. *Critical Public Health, 18*, 5–20.

Fayerweather, W. E., Higginson, J., et al. (1991). Ethics in epidemiology. Industrial Epidemiology Forum´s Conference. *Journal of Clinical Epidemiology, 41*(1), 41S–50S.

Gelsinger, P. & Shamoo, A. E. (2008). Eight years after Jesse´s death, are human research subjects any safer? *The Hastings Center Report, 38*(2), 25–29.

Kass, N. E. (2001). An ethics framework for public health. *American Journal of Public Health, 91*(11), 1776–1782.

Kass, N. E. (2004). Public health ethics: From foundations and frameworks to justice and global public health. *The Journal of Law, Medicine & Ethics, 32*(2), 232–242. 190.

Khoury, M. J., Gwinn, M., et al. (2010). The emergence of translational epidemiology: From scientific discovery to population health impact. *American Journal of Epidemiology, 172*(5), 517–524.

Krarup, T. M., & Blok, A. (2011). Unfolding the social: Quasi-actants, virtual theory, and the new empiricism of Bruno Latour. *The Sociological Review, 59*, 42–63.

Landry, R., Amara, N., et al. (2006). The knowledge-value chain: A conceptual framework for knowledge translation in health. *Bulletin of the World Health Organization, 84*(8), 597–602.

Link, B. G. (2008). Epidemiological sociology and the social shaping of population health. *Journal of Health and Social Behavior, 49*(4), 367–384.

Macklin, R. (2004). *Double standards in medical research in developing countries.* Cambridge, New York: Cambridge University Press.

Madon, T., Hofman, K. J., et al. (2007). Public health. Implementation science. *Science, 318*(5857), 1728–1729.

Petryna, A. (2007). Clinical trials offshored: On private sector science and public health. *BioSocieties, 2*, 21–40.

Porter, D. (2005). *Health, civilization and the State.* London, New York: Routledge.

Victora, C. G., Habicht, J. P., et al. (2004). Evidence-based public health: Moving beyond randomized trials. *American Journal of Public Health, 94*(3), 400–405.

Weed, D. L. (1994). Science, ethics guidelines, and advocacy in epidemiology. *Annals of Epidemiology, 4*(2), 166–171.

Weed, D. L., & McKeown, R. E. (1998). Epidemiology and virtue ethics. *International Journal of Epidemiology, 27*(3), 343–348. discussion 348–349.

Weed, D. L., & McKeown, R. E. (2003). Science and social responsibility in public health. *Environmental Health Perspectives, 111*(14), 1804–1808.

Wilkinson, R. G. (1994). The epidemiological transition: From material scarcity to social disadvantage. *Daedalus, 123*, 61–77.

Winkelstein, W., Jr. (1996). Eras, paradigms, and the future of epidemiology. *American Journal of Public Health, 86*(5), 621–622.

Winters, D. J. (2006). Expanding global research and development for neglected diseases. *Bulletin of the World Health Organization, 84*(5), 414–416.

Chapter 11
Integrating Bioethics in Public Health

Abstract Acknowledging world-wide socioeconomic inequalities and health disparities has inspired a call for global ethics, but *Realpolitik* remains oblivious and continues to support economic and political globalization that weakens nations, increases inequity, and condemns the poor and disempowered to helplessness. International proclamations notwithstanding, actual commitment to improve the social, sanitary, and environmental problems increasingly plaguing the less fortunate majority of the world's population, are feeble or absent. These conditions are deepened and entrenched by *status quo* politics, to the point that poverty, need, and disparity are being naturalized by those basking in the benefits of technoscientific progress and material affluence.

Academic support of global justice and universal human rights has been ineffective. The ethics of protection strongly endorses public health practices that are true to the democratic mandate of engaging in bottom-up acknowledgement of the needs burdening vast segments of human population, as well as appreciable pockets of marginalization and poverty in the backyard of the affluent. If protection takes priority over justice, as here proposed, public health will give utmost importance to the most basic medical needs, and apply its efforts to the immediately and deeply health-damaging environmental disorders. Guided by the ethics of protection, public health validates its claim to eliminate health inequalities and reduce health-related social and environmental factors that hamper people's empowerment.

Keywords Democracy • Health care equity • Naturalization • Participation • Realpolitik • Socioeconomic determinants • Upstream policies

Historical processes mark the politics and the ethics of a nation. Colonialism leaves sequels and perpetuates inequalities; nations long immersed in war and terrorism tend to burden public health with defensive and retaliatory tasks, including medical tolerance and support of torture (Gross 2006). Nations that have been guilty of genocide will show extreme reticence to discuss euthanasia, while societies that have suffered oppression and dictatorship tend to fortify democracy in order to avoid relapses, and correct the iniquities that tyranny has perpetuated. Cultural diversity is a potent obstacle to the idea of global ethics or the universal and sturdy recognition of human rights.

Quandaries of Globalism

Almost 60 years after the "Universal Declaration on Human Rights" and the recent "Declaration of Bioethics and Human Rights" (2005), the outlook for human prosperity looks bleaker than ever. One fifth of developed countries use up to ¾ of the world's energy: a "typical citizen" in more developed countries consumes in 6 months what an inhabitant of less developed countries will access during a whole life-time (Harper 2004). Food and water supply, and of course health care, follow similar patterns of inequality, talk about global justice losing all credibility and raising the question who is being addressed by statements like "Good global governance is crucial" (Friel et al. 2008). Holistic demands for global health inequality converge with those for sustainable development having in mind the environmental problems of climate change and biodiversity reduction. The Commission on Social Determinants of Health suggests becoming active at three fronts: achievement of global health equity, eradication of poverty, and climate stabilization; yet, lacking a specific addressee, such pleas remain unheeded. These major tasks will only be broached if disaggregated into local and specific actions.

The more grandiose plans for global justice and health equity collapse, and commitments made at Alma Ata, Doha, Kyoto, Copenhagen, or Cancun remain unfulfilled promises. From one international conference to the next, participants acknowledge failures to reach established goals, while doting on triumphs in restricted areas of intervention that may show measurable improvements but are unable to secure their own durability, or further extend their success on a larger scale. Complacent celebration of these events and the documents they produce go in the wrong direction, for they lull initiatives for reform and improvement.

Philosophical agendas for social justice and concern for future generations are unbound ethical propositions strongly suggestive of utopia. In fact, they conceive circumstances that will never obtain in real life, they have no real or potential *topos*, hence their u-topic character. Utopia is dangerous, it deflect from present realities and shortcomings, distracting from recognition of the scarcities and unhappiness that we are advised to tolerate as the price to be paid in the name of a glorious future.

Comprehensive justice theories, whether based on material redistribution or focused on political rights, are top-down propositions that rely on the supposedly precise diagnosis of what is needed. However, downstream need theories fail to correctly assess the starkness of deprivation. Theoretical listings of health care needs tend to be paternalistic, systematically neglecting the actual deficiencies felt and indigence suffered by vast populations.

Public health cannot tackle global socioeconomic determinants. It acknowledges the impact of social causation on health inequalities, but will only be efficient when concentrating on health care policies and disaggregating them from influential determinants it does not have the power to seize. For the same reason, public health programs should not be seduced by global schemes that provide earmarked assistance at the cost of neglecting national commitments.

Attending the needs of the destitute and unprotected brings public health to a welcome nearness of processes of empowerment and the development of capabilities to engage in deliberation and participation. Basic needs must be recognized, thus inevitably giving precedence to public health programs that are proximal and extensive to the population that depends on the governance of the State. Global ethics can only gain relevance if it supports and reinforces national tasks and programs, which is precisely the mandate that global forces have been unwilling to adopt. Specific problems may need a still more limited field of action, but on occasions transnational efforts will be required when pandemics threaten. Even here, international cooperation does not go beyond interest-protecting negotiations. The WHO recently came up with a "pandemic influenza preparedness (PIP) framework," hailed as a "milestone in global governance for health." In fact, the agreement invites member States to share viruses with human pandemic potential and genetic sequences, to be studied by a WHO Global Influenza Surveillance and Response System, co-financed by private laboratories and vaccine manufacturers. Inauspiciously, resulting intellectual property right disputes were not solved, and no "norms encouraging developed countries to make specific equity-enhancing contributions to developing countries, such as donating portions of purchased vaccine" were proposed (Fidler and Gostin 2011).

Perils of Naturalism

The distinction between natural and anthropogenic realities, and the dichotomy between *natura/cultura* or nature and nurture—a duality introduced by Galton who favored natural traits over acquired ones—tends to blur or be manipulated. Both present day socioeconomic reality and the idea of future generations are often seen as natural situations and processes, falling prey to sociologist W.I. Thomas' theorem: "if men define situations as real, they are real in their consequences" (Merton 1995).

Tendencies to preserve a social *status quo* are based on the tenet that socioeconomic conditions are so deeply entrenched as to be accepted as natural unalterable realities immune to strategies bent on change or regulations. Naturalizing historical and social processes is a form of domination that discourages any initiative to improve the lot of the worst off and put them on their way toward basic security and empowerment (Schrecker 2008). Claims for justice and respect of human rights lack the force to address controversies between social weal and individual autonomy, repeatedly showing their inability to defuse conflicts of power.

Conservative views tend to naturalize social phenomena, ecologic deterioration, power gradients, and market laws, unraveling ideas about human nature as given and unwieldy to change. Poverty has also received the tag of a natural status that should not be assisted but let to dwindle into spontaneous extinction, as suggested by Maltus, Galton, Hardin, and others. Naturalism in its most perverse form is at the basis of racism, homophobia and other forms of violent intolerance. Possibly, the hardest naturalistic nut to crack by public health is the idea of "determinants" as contrasted with "modifiable" risk factors amenable to intervention (Last 2001, quoted in Kindig 2007).

Technoscience colonizes natural processes arguing that it has been the mark of mankind to interfere with nature and adapt it to human needs and desires. Public health's concerns go in the opposite direction, unwilling to see anthropogenic realities as frozen into natural, unchangeable events and processes. So-called socioeconomic determinants must be recognized as anthropogenic and therefore amenable to reformulation and modification. Bioethics has reflected upon man-made transformations of natural traits such as crop genomics, reproductive cloning, or the intervention in ecological niches. Furthermore, progress promises to solve the problems it creates, but evidence points to developments where solutions lag behind side-effects, and the benefit/costs balance favors the privileged at the expense of the less fortunate.

Conservative politics and the unrelenting realizations of technoscience tend to settle man-made artifice as naturally given and to accept its impact on human biology. Health care and disease management must adapt to solidly anchored externalities, and public health is expected to cooperate by turning its gaze toward the individual and help him adapt to a world immersed in processes that cannot be subjected to rational or ethical control. Inasmuch as epidemiology continues to confirm the solidity of socioeconomic determinants, it delays all efforts to neutralize external health threatening risk factors, all the more so if global ethics is broadly called on to address these quandaries.

Social epidemiology and molecular epidemiology press public health into naturalistic thinking, the first one by giving socioeconomic determinants an almost ontological status, the second by establishing genetic determinism in health and disease. Following both perspectives, it can be postulated that genes will determine the way socioeconomic factors will be influential, just as socioeconomic conditions will determine which genetic traits shall be expressed, modulated, or suppressed. In any event, deterministic forces would discourage public health intervention and deprive it from cutting edge influence, if they really told the whole story. Fortunately, we are reminded that much can be gained by recovering an interest in "cultural, historical, political, and other population factors" (Pearce 1996). And, it might be added, substantively enriching the discipline of public health with a robust bioethical perspective.

Public Health and Democracy

In 1849 R. Virchow proclaimed that "a *reasonable State Constitution must unmistakably establish the individual right to a healthy existence*" (cited in Deppe and Regus 1975, italics in original). Nearly two centuries later, political philosophies have failed to find a stable balance between individual existence and fair social orders, either privileging socialism at the cost of freedom, or celebrating liberty at the expenses of social cohesion and solidarity.

Public health appears to be in a quandary: the very poor cannot be helped because resources are scarce and progressively restricted by global policies that bet on economic success at the cost of social commitment; and the better off are expected to take care of themselves by subscribing to private initiatives providing social security and medical services to those who can afford them. The net effect is that health inequalities are on the rise in spite of good intentions proclaimed by the academic world and international institutions. Statistics on poverty, inequality, scarcity, and environmental decay are alarming, while social security plans in welfare States erode and deteriorate. Citizenries are unable to gain a firm hold on democratic participation, let alone decision-making empowerment, and the voiceless are bereft of hope. The trend of world economics anticipates that wealth will continue to concentrate, while the costs of social services—especially health and medical care—will go on spiraling, driving vast population segments to remain unattended, precariously insured, and compelled to copayment for services no longer rendered by the State.

Beyond such exceptions as pandemics, catastrophes and world-wide crises, public health has to deal nationally with the influence of global socioeconomic factors that are utterly beyond its reach of influence. This leads to a very uneasy fit between theory and reality, blunting the edge of some fine bioethical arguments that lack practical purchase, often reaching for utopia and losing touch with reality. Defenders of fair resources allocation expand their agenda to include such abstract primary goods as "social basis of self-respect"; capability theorists list intangibles like being happy and able to have pleasurable experiences including opportunities for sexual satisfaction, as part of a long, admittedly open-ended, list (Nussbaum 2006).Whom are these philosophers talking to? There is no instance, social, political, or economic in the realm of *Realpolitik* that will seriously consider an open-ended quest for justice, and democratic participation in public health ventures fails to get a hold on such extravagantly formulated goals. Inapplicability may not be a good argument against an ethical proposal, but it lacks leverage for applied ethics.

Public health ethics has to consider all these aspects and dismiss the ingrained tendency of ethics to present universalizable principles (Hare 1981). At the same time, fragmentation and provincialism in ethical reasoning must be avoided, for unless applied ethics shows coherence and a tendency toward generalization it will be unreliable, inconsistent, and of little guidance to policy-makers and practitioners. Values and preferences must be clearly articulated, upheld, and defended, at the same time exercising tolerance and acceptance of different convictions provided they are intelligibly explained and consistently practiced. Democracy and pluralism require all actors and viewpoints to engage in mutual recognition, cultivating respect and rejecting dogmatism or authoritarianism that might interfere with the liberty of friends and strangers, provided all hold to the basic ethical maxim of rejecting harm. Values and principles need to be resistant to erosion and change, but also flexible enough to share or cede influence. Deliberation is to remain actively persistent, for changing circumstances and ongoing developments demand reappraisals and *prima facie* reordering of priorities.

The unavoidable tension created by public policies that inevitably impinge on individual autonomy is innate to, and especially profound in, sanitary matters where generalized norms affect human beings in their biological singularity. Ethical justification for public health actions must rely on the premise that public health promotes the common good by efficiently advancing shared interests that will potentially benefit everyone.

Insecurities and underdeterminations breed opportunities for manipulation, as well as the unbridled expansion and consolidation of vested interests. Democracies that place great value on participative deliberation require that uncertainties be openly discussed instead of leading to authoritative top-down decisions. The subjective perception of risks remains valid as long as objective evaluation of threats are wanting; in the absence of scientific information, public opinion formed through participatory deliberation prevails as the only valid criteria to evaluate whether action or procrastination will better serve the need of protection.

Should the New Public Health movement continue to prosper, public health in its traditional sense will become restricted to a few hygienic and sanitary interventions, and the control of transmissible diseases. The New Public Health betrays the original social dedication of public health, deploying it in a direction that increases health inequities, especially in regions that already suffer from severe social injustice. Public health is under constant pressure to privatize medical care, lodging strategies of individual preventive actions in the clinical encounter. Social injustice, massive poverty, and progressive deterioration of the middle classes will remain entrenched beyond reform unless protection becomes the leading inspiration of public health. As the amount of unsatisfied needs increases and social security shrivels, it becomes imperative that scarce fiscal resources be first assigned where most pressingly needed.

Ethical approaches to public health in the form of prevention and precaution have been unable to legitimate personal adherence to public interventions; quite to the contrary, social security networks have been dismantled in favor of private initiatives catering to paying individuals. The ethics of protection in public health adhere to unconditionally safeguarding health and medical needs through national services that help establish a social capital based on citizens' morality of cooperation and solidarity. Public health actions must be effective and relevant enough to legitimately require citizens to do their due by accepting certain restrictions of autonomy, and acknowledge that individualism can only flourish in the wake of a robust common weal that protects and empowers all its members. To this end, protective bioethics will support public health measures when sufficient evidence justifies effective intervention and disciplined individual participation. Protection should take preeminence over other ethical theories whenever these conditions obtain.

Outlook

Public health analysts are often tempted to give their views on the future of the discipline in a mixture of sober prognostication and wishful thinking. Reducing health disparities ranks high on the list of future challenges, as does the framework of health-care systems that hope to balance the undisputed triad of equity, cost, and quality, even though admitting that they may be incompatible, for quality has costs, and costs put equity under tension. Other challenges are debatable, like increasing life-expectancy which is fine provided old-age is socially protected, or addressing issues which public health is unable to cope with, as violence, pollution, undesirable life styles, and complex social interactions. Dealing with genetic diseases, new infectious conditions, and mental health disorders will probably continue to fall in the resort of clinical medicine and be available under market conditions with little or no public health participation (Koplan and Fleming 2000). Other challenges such as enhancement biology, the technical development of cyborgs and posthuman beings, the potentials of nanotechnology and the neurosciences, and their impact on our concepts of normality, health, and disease, are as yet dimly visualized possibilities.

The complexities of modern society and the expanding intricacies of world affairs favor expert focal knowledge but work against a comprehensive, well-substantiated grasp of major problems. History offers convincing examples of effective public health campaigns when based on clear and undisputed evidence: citric food to prevent scurvy, disinfection of milk through pasteurization, universal vaccination against small pox, were all successful campaigns based on cause–effect knowledge. Compared to these triumphs, the management of HIV/AIDS, influenza epidemics and pandemics, and nontransmissible diseases are challenges steeped in uncertainty, rendering public health powerless. Epidemiology, at present seemingly lost in a world of research efforts that either fail to offer pragmatic solutions or take refuge in long-range promises of improbable fulfillment, must revert to a democratically inspired science responsive to the actual problems that public health faces and needs to confront.

The most plausible and down to earth anticipation is to take care of actual problems, and rest assured that their solution will benefit present and future lives. The only fair and consistent way public health can fulfill its tasks is by assuming both health and medical care for the deprived *and* the reduction of exposure to external risks. In both instances, if protection takes priority over justice, public health will attend the most basic medical needs first, and also give priority to the detection and neutralization of the most immediately and deeply health damaging environmental disorders.

Under the guidance of an ethics of protection, public health validates its claim to engage in efforts to reduce health inequalities and take action to eliminate health-related factors that hamper people's empowerment.

References

Deppe, H.-U., & Regus, M. (1975). *Seminar: Medizin, Gesellschaft, Geschichte*. Suhrkamp Verlag: Frankfurt a.M.

Fidler, D. P., & Gostin, L. O. (2011). The WHO pandemic influenza preparedness framework: A milestone in global governance for health. *JAMA, 306*(2), 200–201.

Friel, S., Marmot, M., McMichael, A. J., Kjellstrom, & T., Vågerö, D. (2008). Global health equity and climate stabilisation: A common agenda. *Lancet, 372*(9650), 1677–1683.

Gross, M. L. (2006). *Bioethics and armed conflict*. Cambridge, London: The MIT Press.

Hare, R. M. (1981). *Moral thinking. Its levels, method and point.* Oxford: Clarendon Press.

Harper, C. L. (2004). *Environment and society.* Upper Saddle River: Perason Prentice Hall.

Kindig, D. A. (2007). Understanding population health terminology. *The Milbank Quarterly, 85*(1), 139–161.

Koplan, J. P., & Fleming, D. W. (2000). Current and future public health challenges. *JAMA, 284*(13), 1696–1698.

Merton, R. K. (1995). The Thomas theorem and the Matthew effect. *Social Forces, 42,* 379–424.

Nussbaum, M. C. (2006). *Frontiers of justice.* Cambridge, London: TheBelknap Press of Harvard University Press.

Pearce, N. (1996). Traditional epidemiology, modern epidemiology, and public health. *American Journal of Public Health, 86*(5), 678–683.

Schrecker, T. (2008). Denaturalizing scarcity: A strategy of enquiry for public- health ethics. *Bulletin of the World Health Organization, 86*(8), 600–605.